By Margie Kay, Wayne Lawrence, and Bill Spicer

The Fast Movers:

High-Speed UFOs/UAPs

THE FAST MOVERS:
HIGH-SPEED UFOS/UAPS

By Margie Kay, Wayne Lawrence, and Bill Spicer

Edited by Margie Kay

Published in the United States of America

ISBN: 978-0-9988558-6-8

Library of Congress Copyright TXu002196216

Published by Un-X Media
P.O. Box 1166
Independence, MO 64051
www.unmedia.com

Bulk copies may be purchased from the publisher at a discounted rate.
Please send an email to editor@unxmedia.com for more information.

UNXMEDIA

PUBLISHING

Acknowledgements

Thanks to Bill Spicer for discovering the Quantum UFO Observation Technique on his own and sharing it with all of us.

Thanks to Wayne Lawrence for his tireless efforts taking video and examining hours of film footage to find the Fast Movers.

Special thanks to Kristina McPheeters for her excellent illustrations.

Thanks to Kerry Walker, Tony Degn, and Gary Sansbury for sharing their photographs of fast movers.

Thank you to the Mutual UFO Network for providing a format for people to file UFO reports

"*If you want to find the secrets of the universe, think in terms of energy, frequency, and vibration.*"

Nicola Tessla

TABLE OF CONTENTS

A high-speed UFO captured on film using the
Quantum UFO Observation Technique

Photo: Bill Spicer

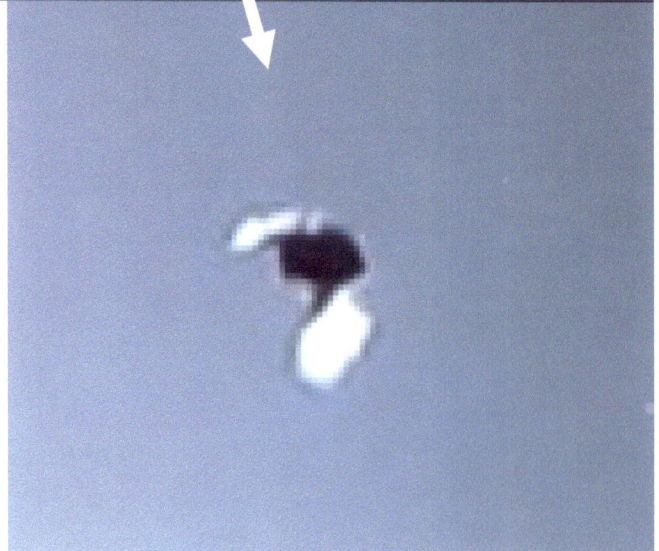

INTRODUCTION

This book is a combined effort of three individuals with different perspectives who describe the use of different modalities to capture high-speed UFO/UAP phenomena.

I have personally observed many fast-moving craft in our skies using the naked eye and by remote viewing —but most people never notice these craft, and certainly most never notice them in videos, either. This is because the Fast-Movers, as we have coined them, move so fast that they can usually only be seen when looking at video frame-by frame, or examining individual photos snapped at high speed. Only then it becomes apparent that there are many unknown craft in the skies over the Earth.

I asked two long-time friends and colleges in the field of Ufology - Bill Spicer and Wayne Lawrence - to collaborate with me on this book because I knew that each of us had a unique perspective of the situation: One quantum telepathic approach (mine), one scientific approach (Wayne's), and one doing both (Bill).

In order to fully understand the Fast Movers, a person must realize the fact that we live in a multi -dimensional universe and that in order to consciously interact with it, an individual must first have an open mind, and second, wake up and learn to observe what is going on all around us on a constant basis.

We are not alone, as they say.

This book, if read in its entirety, will give anyone the foundation needed to become aware of, and not only observe the Fast-Movers themselves but learn techniques for seeing them with the naked eye and capturing them on film.

I urge the reader to try some of the techniques we outline in the forthcoming pages and see the amazing results that can be obtained.

Margie Kay

CHAPTER 1

FAST MOVERS CAPTURED ON VIDEO

FROM THE PERSPECTIVE OF A MUFON FIELD INVESTIGATOR

By Wayne Lawrence

I began to investigate my own sightings of Unknown UAVs (Unknown Aerial Vehicles) in 2014 when I began to see them. I did not discover fast movers until late August of that year. As of September 2019, 16 of my 39 personal sightings contained Fast Moving craft. That represents over 38% of my sightings.

In addition, as of December 2019, I have investigated 74 MUFON reports that have been assigned to me and in only four of them there were Fast Movers. The disparity in the numbers comes from the fact that I nearly always take video of my sightings, and most of them are in daylight. The only three MUFON cases that submitted daytime video had Fast Movers. Many of the reports that come to MUFON only have night time videos, because that is when people notice them. I also find that daytime reports are often made when witnesses are driving or don't have a camera handy.

A total of 21 cases that I have investigated contain as many as 90 Fast Movers. That averages out to over 4 per case. The following analyses and images are from those investigations. I have presented only the best of the images and not all of them. The methods of obtaining those images revolve around the following:

Quantum UFO Observation Technique (Shooting photos with filters.)

Random daytime videos with a Nikon D-3300 camera plus 300 mm zoom.

Random daytime videos with a Full Spectrum adapted video camera.

Video or still images in which the focal point was backed out to enlarge the image.

Smartphone video and still images.

I was introduced to the Quantum UFO Observation Technique by Bill Spicer in August 2014. It has proved to be a useful method in catching fast movers although it seems that they are usually quite a bit further away and difficult to focus on since one cannot see them in order to focus adequately.

All the other types of images were discovered as I looked over videos with frame by frame analysis. There are only about 3 exceptions in which I accidentally caught Fast Movers in still photos that I had shot.

I am amazed at the number of these rapidly moving craft. There are some very surprising speeds that are observed as well. The huge variety of these types of UAVs is important to highlight. This variety of shapes also is true in classic types of UFOs that are reported. One

should not be surprised to find that it is true with Fast Movers as well. In fact, I would rather expect that extraterrestrials observing our planet would also note that there are many different types of vehicles that earthlings drive, sail, or fly. To each their own!

Enjoy the investigations. Perhaps someday you too may be able to catch some of these Fasts Movers!

Many Fast Moving Craft

On Friday, August 29, 2014 I decided to shoot a video at 6:21 pm in the western sky at an angle of elevation at about 20 degrees above the horizon. I shot the video using the technique of shooting against a very small portion of the Sun, called Quantum UFO Observation Technique (QUOT). Most of the Sun is blocked by the roof of a house. A polarizing filter and UV filter are attached to the lens of the camera. The brightness of the small piece of sun causes

the chip in the camera to be sensitive to bright objects that would not normally be caught.

Different shapes moved rapidly across the screen. Some were like a satellite, some were ghosted spheres, others were dark spheres, others constantly changed morphologically. The image below shows the first frame of the video sequence with a sliver of the sun blocked by the roof.

These "satellite-like" craft flew across the screen in $3/30^{th}$ of a second, or .1 seconds. As one can see, it was in the process of closing the wings and pulling them into the body of the craft. Note also the fact that the rear of the craft is wider than the front, which aerodynamically would be very difficult to manage if it was following normal wind dynamics.

The following are examples of some of the craft that move very rapidly in the video:

Frame 0:26

Frame 0:25

Frame 0:24

Snapshot 15

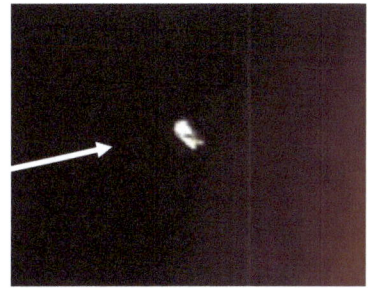

Detail of craft

Snap Shot 17

Large grey sphere

Small white sphere

Snapshot 20: and at right—sphere enhanced and enlarged

Snapshot 141 taken 10/27/14: and at right—object enhanced and enlarged

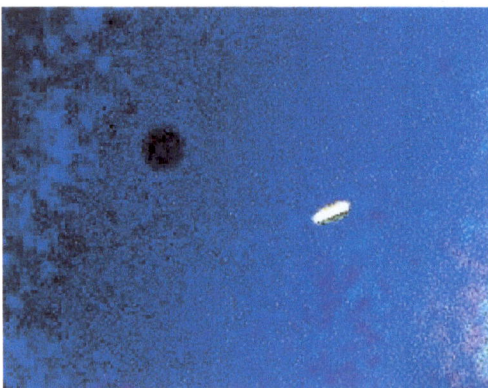

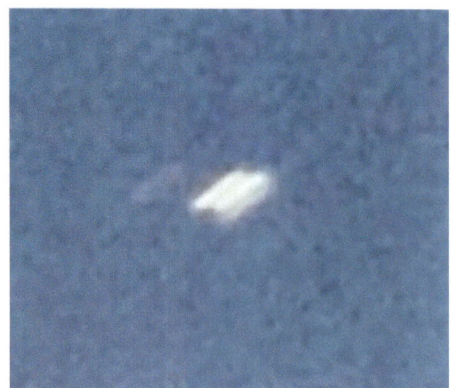

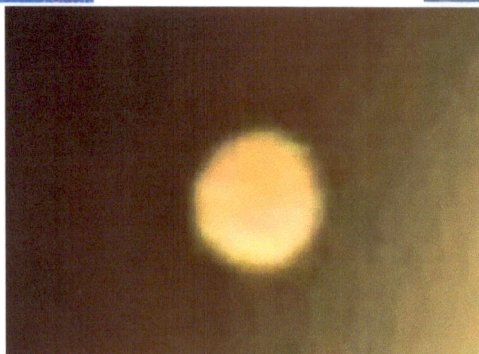

I took many videos by shooting against the Sun in this manner. Not all of the images turned out very well, but one can definitely see that there is an object moving very rapidly. These above are some of the clearer ones that I was able to capture and process.

Note: In the "Satellite-like" object above, the question was raised about whether it could possibly have been an insect. Insects have a body comprising of three sections: the head, the thorax, and the abdomen. On the head are various parts including antennae, mouth with pincers, eyes, and sometimes other appendages. The thorax has 6 legs hanging down. The abdomen is pronounced. On the top of the thorax are wings, one per side. Clearly that was no insect.

Summary:

This report highlights 14 different types of craft that were seen primarily on Friday Aug. 29, 2014 although one was from Monday Oct. 27, 2014. None of them were seen with the naked eye but only with frame by frame analysis. Many images have been videotaped using this method however due to the speed by which they travel, most images are too blurred to see definite shapes. These Fast Movers are pervasive above the skies and can be seen virtually anytime one wants to set up a camera.

Black sphere, 2 rods

Two objects

Five objects in one day

MUFON Case # 60295

This case was to be the second time that I noticed that there were some very fast-moving craft moving in the videos. They could not be seen simply by viewing the video. Here is the context:

On Sunday September 21, 2014 I had begun taking some photos at different times due to the clear sky at the time. I had been noticing odd things… like momentary flashes of craft but I wasn't totally sure if I was just seeing things. So, **at random**, at 11:09 am, I decided to just point and shoot and take a video in an area that I had seen craft at for sure earlier. I didn't see anything of interest upon viewing it at first.

Craft #1--At 11:30 a.m., I noticed a **white object** above one of the trees in the back yard which suddenly appeared and seemed to be staying roughly in the same place. I took a series of five still shots on my 300 mm zoom and the focus was set near infinity. The direction was almost due north for the first one, and about 355° for the fifth one. I knew it was not a bird or plane or anything identifiable. The craft appeared to actually be two of them in tandem, as it bears some resemblance to the white craft in case # 59835 (seven days earlier). It appeared to have been hovering over the Blue Springs Lake area at about 1000 ft, and 15° above ground level.

Craft # 1, not a fast mover

When I took the first video at random, I did not see anything at first. Then with frame by frame analysis, I began to see very slight movement. I say slight because they were light gray against the sky and hard to see until I knew what to look for. The important thing is that these faint grey craft seemed to be coming from the Blue Springs Lake area, rising at extremely rapid rates, and occurring quicker than the human eye can see. In both movie sequences, they rose and moved out of the scene area in 1.23 seconds or less, some in as little as .26 seconds. Not only that, but craft #2 described below made sudden turns that are beyond any human technology. The "g" forces alone would be incredibly high unless accounted for in some way.

Craft #2 Zigzag—In the first video sequence—the craft shot up at an angle of 25° and the 37 frames elapsed in 1.23 seconds. The craft travelled straight, effected **a sudden 90° turn**, three frames later **a sudden 45°** turn and left the screen. This zigzag pattern was remarkable. **The speed was 205 mph.**

Craft #3—Second movie sequence--shot up at an angle of 40° and the 20 frames of its observed flight elapsed in .66 seconds. **The speed was 196 mph.**

Craft #4—Second movie sequence--shot up at an angle of 55° and the 8 frames of its observed flight elapsed in .26 seconds. **The speed was 340 mph**.

Craft #5—can barely be seen in the photomerge image below next to the arrow.

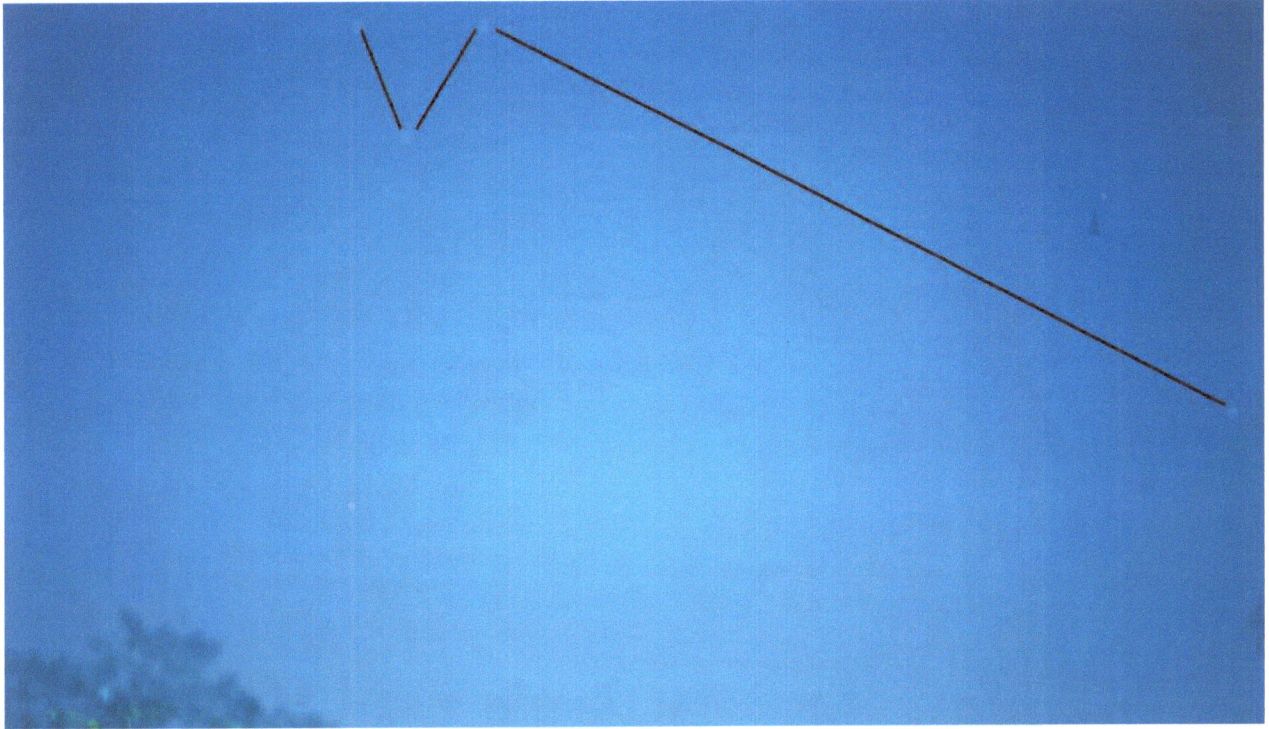

Photomerge of craft #2 grey sphere movements from first movie sequence.

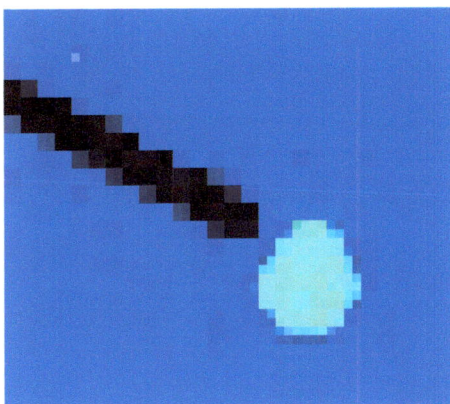

Craft #2

Craft #3 edited

Craft #4 edited

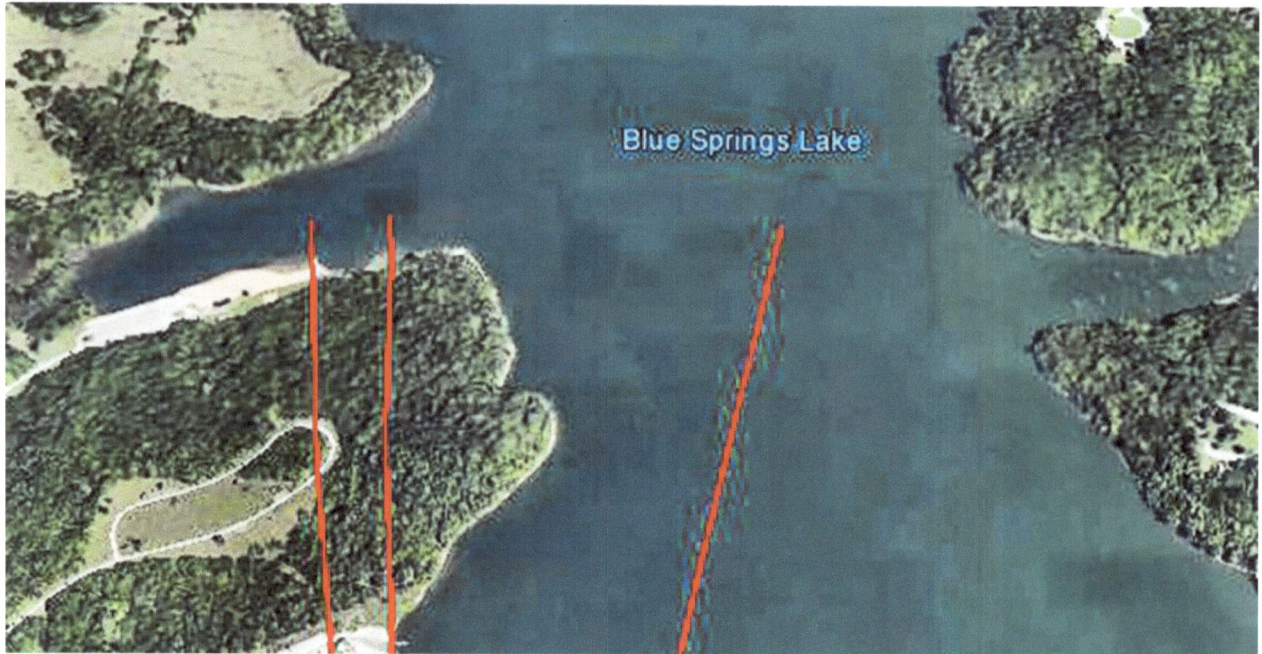

This **satellite photo** on the prior page shows dark areas in Blue Springs Lake. They appear to be **rectangular**, which is odd for the smooth bottom of a smooth lake. (The two red lines close to each other point to the area covered by the zigzag). The far right red line points to where a direct line reverse trajectory indicates direction of origin of craft #2. Summary: Clearly these fast movers are well named!

Very Fast Moving Craft

When I took the first video at random, I did not see anything at first. Then with frame by frame analysis, I began to see very slight movement. I say slight because they were light gray against the sky and hard to see until I knew what to look for. The important thing is that these faint grey craft seemed to be coming from the Blue Springs Lake area, rising at extremely rapid rates, and occurring quicker than the human eye can see. In both movie sequences, they rose and moved out of the scene area in 1.23 seconds or less, some in as little as .26 seconds. Not only that, but craft #2 described below made sudden turns that are beyond any human technology. The "g" forces alone would be incredibly high unless accounted for in some way.

Caught on Random Video

MUFON Case # 64198

On Saturday, Jan. 24th, 2015 at 10:45 AM, I had been once again noticing brief flashes of either light, or objects that were momentarily appearing and then disappearing over the Blue Springs Lake area. So, I decided to take some more random video to see if I could catch anything.

I shot a 24 second video in which an object is seen for .53 seconds travelling through the screen and then disappearing below the image. The composite below shows the relationships of the object in the beginning (with the first arrow) and a selection of three other locations as it travelled. The final arrow shows where it left the view on that particular screen. The image below is a photomerge of 5 snapshots from the video.

If we assume that the craft is about 1 mile away, and roughly 15 feet in size, the distance that it travelled downward in vertical distance calculates to be 507 feet. This corresponds to a vertical plunge of 956.6 feet per second, or **652 mph**.

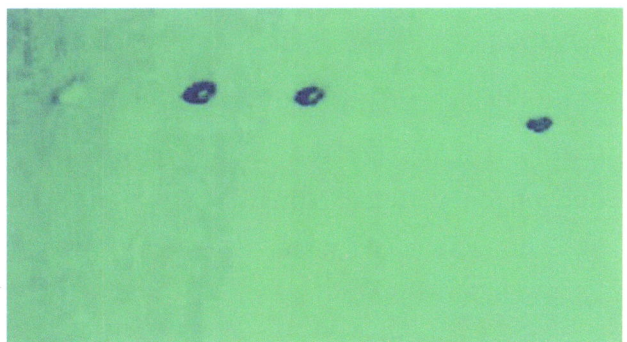

We see frames 1,2,3, and 5 of the sequence in which the first frame is slightly lower than the second frame. This is not a case of freefall, but intentional guidance.

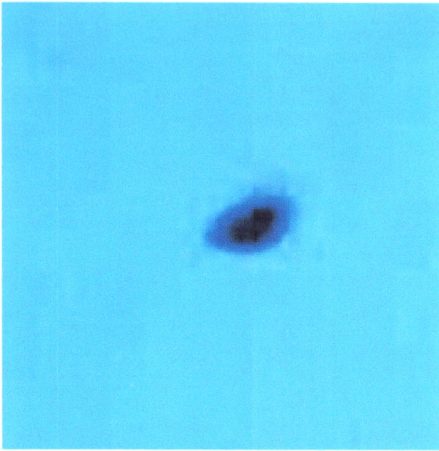

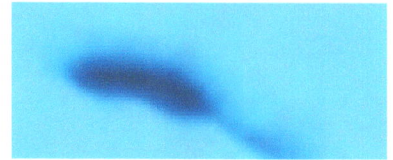

One can see that the craft not only changes its pitch, but also changes morphologically to **three** different shapes all within 1/10th of a second.

The initial direction of the first frame was from 340° or about NNW. When one approximates the continued plunge of the craft, it would have entered Blue Springs Lake at around 5° North. The satellite image below indicates the area at one mile distant from my point of observation.

The plunging craft entered the video at 16° from the horizon, or 1,510 ft in altitude and entered the lake near where many other sightings have been seen.

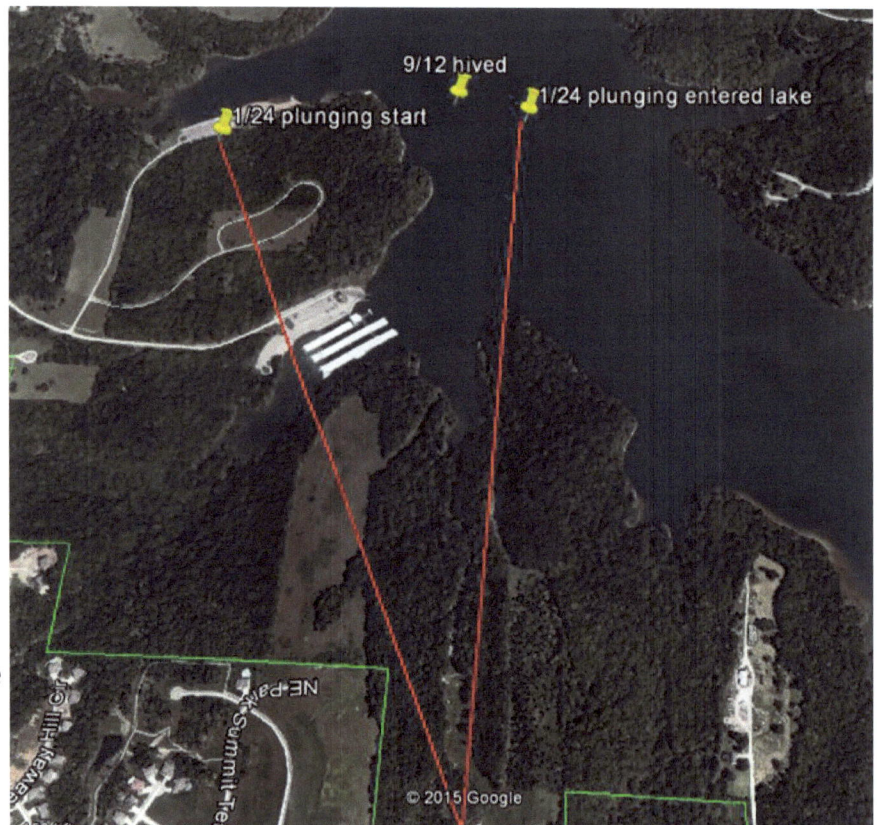

Four on Random Video
MUFON Case # 64199

Sunday March 22, 2015 I had been seeing momentary objects in the air above the woods behind our house. At about 2:15 pm, I decided to take some random videos to see if I could catch anything. In the fourth video, which lasted 18 seconds, I didn't initially see anything on playback. But when I began looking frame by frame, I finally noticed one, and then more objects. The direction was NNW at about 18° altitude.

Object 1:

The first craft appeared as a white object, which changed back to dark and then back several times during the **4 seconds** that it was in view. However, it was faint and hard to detect at first. The following photomerge of four frames during its progress show its path prior to exiting.

Frame :07

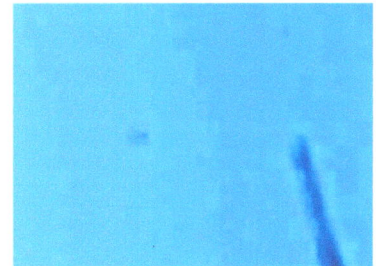

Frame :12

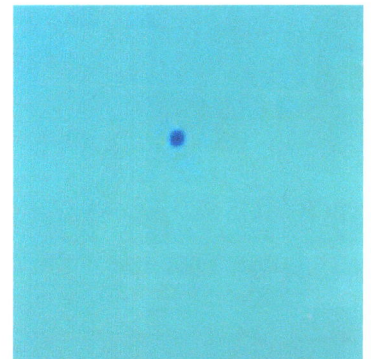

Frame 4.06

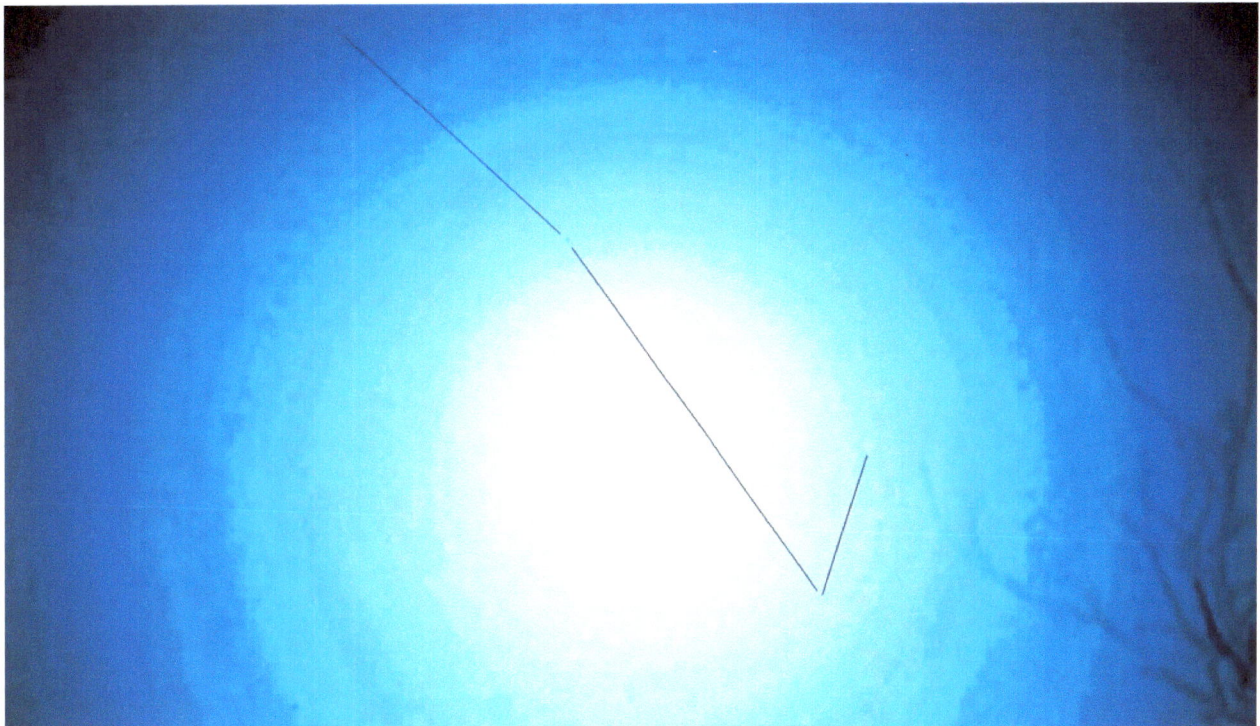

Photomerge of four frames with path drawn in

14

Object 2:

Next, at the 9 second mark into the video another object is seen for only **.83** seconds. Here is it's progress through the sky:

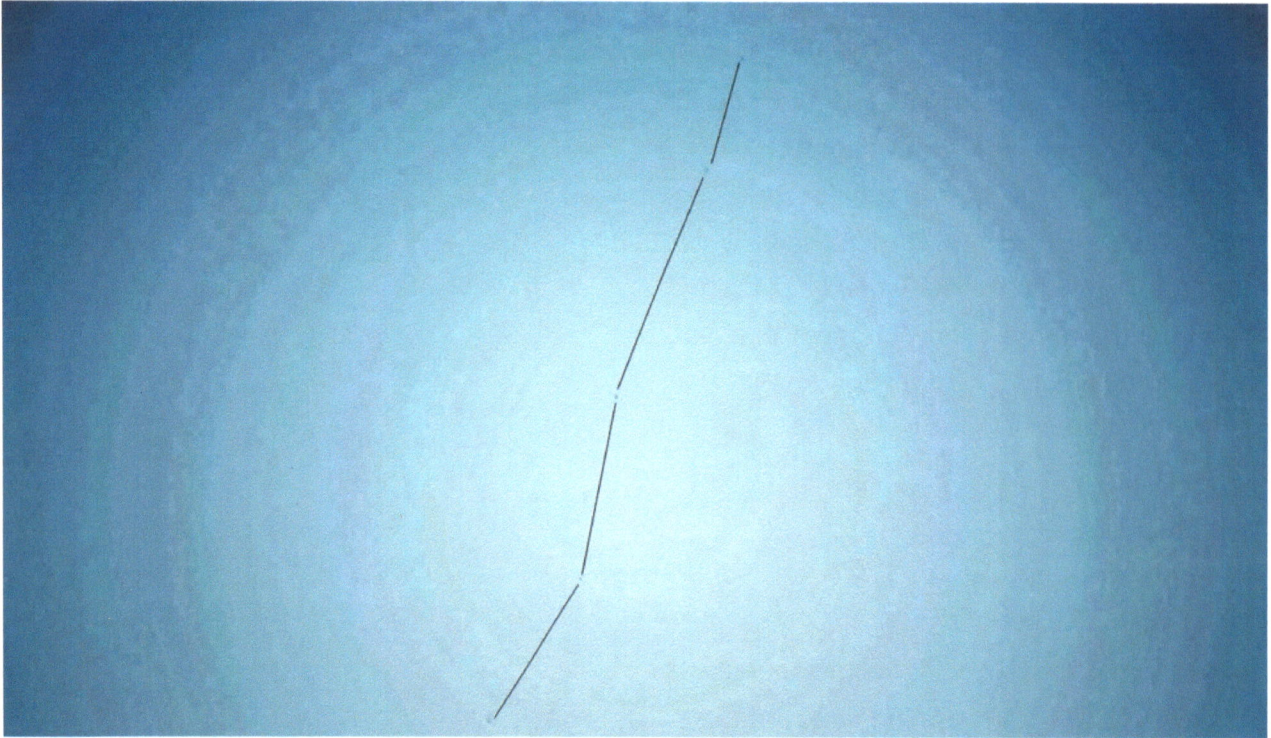

The object itself as it shoots upward shows up as a very small sized craft, though it would be hard to estimate the real size.

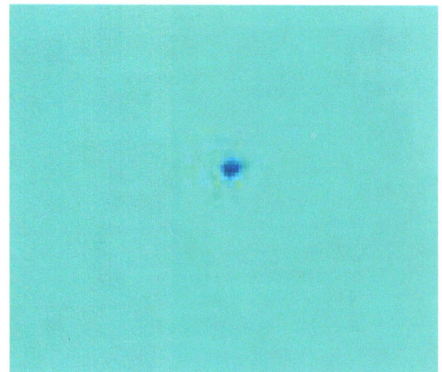

Frame 9:04   Frame 9:14   Frame 9:22

Object # 3:

At the 13 second area of the video, I noticed a craft which lasted for a duration of **.66** seconds. It appeared as a grey or bluish colored object. It was also <u>much larger</u> than the others….perhaps ten times.

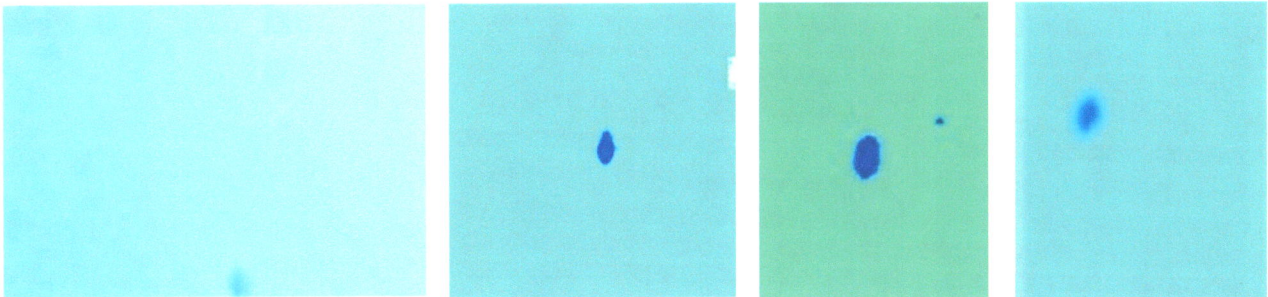

Frame 13:07 emerging from bottom   Frame 13:11   Frame 13:22   Frame 13:27

Interestingly, a smaller object is seen next to it in frame 13:22. Here is what occurred over the course of crossing the sky:

This photomerge depicts five different frames during the upward flight of the object.

16

Object #4:

The final craft that was seen in this video was over a duration of **.7** seconds, covering 21 frames in all. The object appears to be similar in size to craft #2.

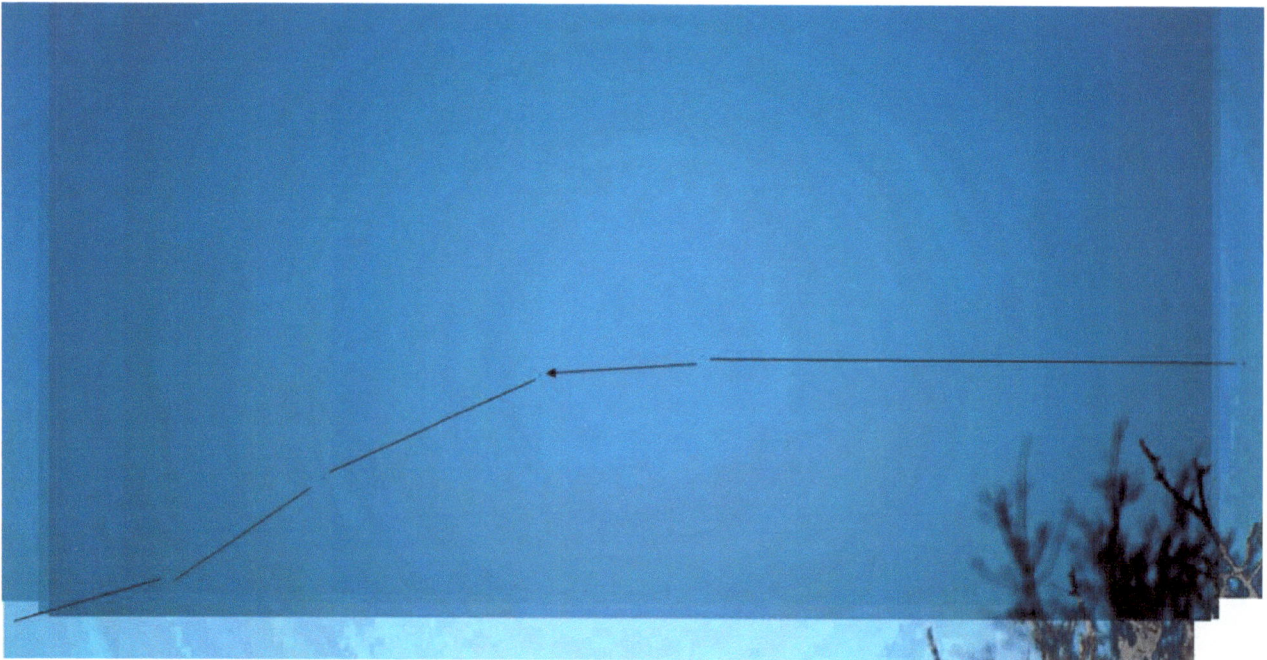

This photomerge is of 5 frames.

The object itself looked like a small round object, just as in number 2.

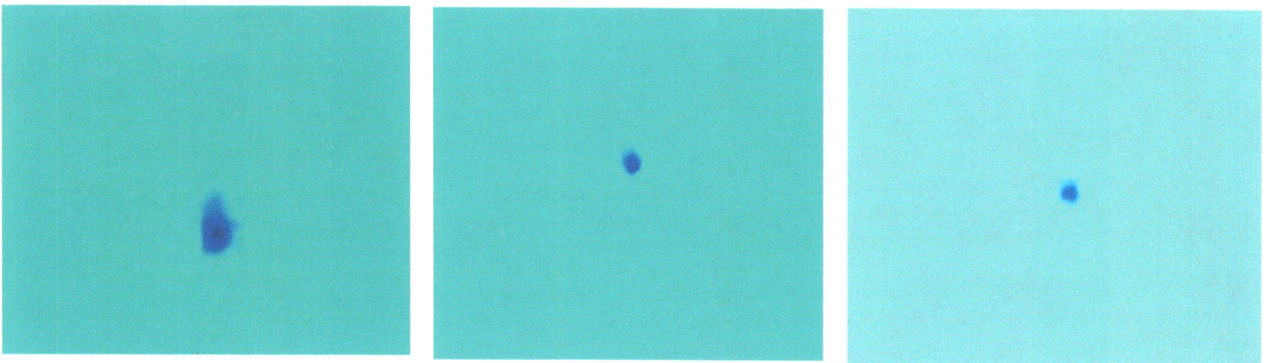

CONCLUSION:

In only 18 seconds of time, there were at least 4 different craft moving faster than the eye can see. I had only noticed momentary flickers of movement which got me to try some random videos. It all has to do with luck, timing, and pointing in the right direction at the right time.

I know that there are many more. This begs the question as to how many of these craft are there in the Blue Springs Lake, and what are they doing?

Jupiter, & Venus Fly-by

MUFON Case # 66969

On the evening of June 8, 2015 at 9:12 pm, I was looking at the planets Jupiter & Venus and decided to take some video with the intention of modifying the focus. My intention was to observe the effects of increasing the size of the image over normal basic image integrity and detail.

Frame 46:34   Jupiter at left, Venus at right

The direction of Jupiter was basically West at 261° and 36° altitude and Venus was at 275° and 29° altitude when I shot 48 seconds of video. I chose Venus due to its greater apparent size.

Upon review of frame by frame images, I discovered that three separate craft flew by the planet Venus from my perspective of view. Here are some images:

Image of Venus

Frame 27:24

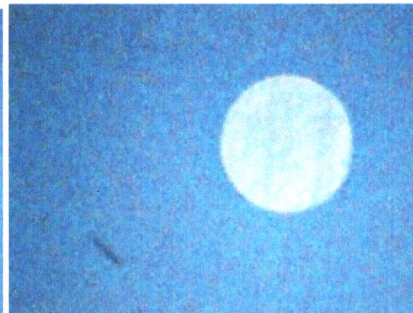

Frame 36:02

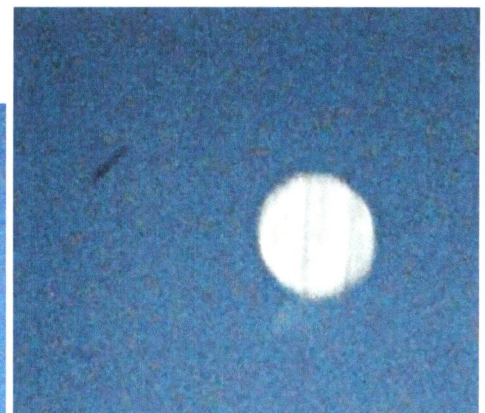

Frame 39:25

Frame 39:25 cropped

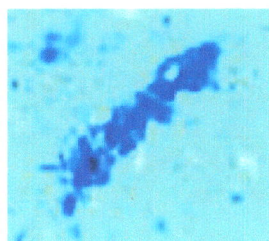

It was pure luck that I even saw the craft. Had I been using a normal focal point, Venus would have been rendered to just a small bright dot and I would not have caught the images of the three craft that flew nearby in the sky. My initial thought was that the distance of the craft from me was 2 to 3 miles.

The first craft flew in a straight line upwards to the left at a 25° angle. The second craft did the same, but at a 50° angle. The third craft actually came from the lower left and did some curving around the planet before continuing on upwards to the right at about a 40° angle. Clearly craft were flying around without lights that evening.

After researching it, the direction of the sightings was toward the Lakewood Lakes. Two of the craft appear to travel from the southern end of the lake and one from the northern side of the most westerly of the two

lakes. The angular size of the craft is .067° compared to expanded Venus which was .2° (unaided at .0067°). The lake is at a distance of 2.2 miles from me and the craft appear to be **13.5 feet** in length, according to my calculations. They could have been closer, or further away. But it is the closest large lake in that area. Most of my observations have been over Blue Springs Lake and Lake Jacomo. Logic tells me that this is occurring here at Lakewood Lakes as well.

I have seen a very large craft on two occasions over Blue Springs Lake that was approximately 117 yds in length. That sighting coincides with the fact that at least two other persons have reported seeing a "football field" sized craft in the Lakewood Lakes area as well. They were reported to have been seen while the individuals were walking along Woods Chapel Road, which is also in the photo below on the next page. Only about 2 miles separate the sightings on the two lakes.

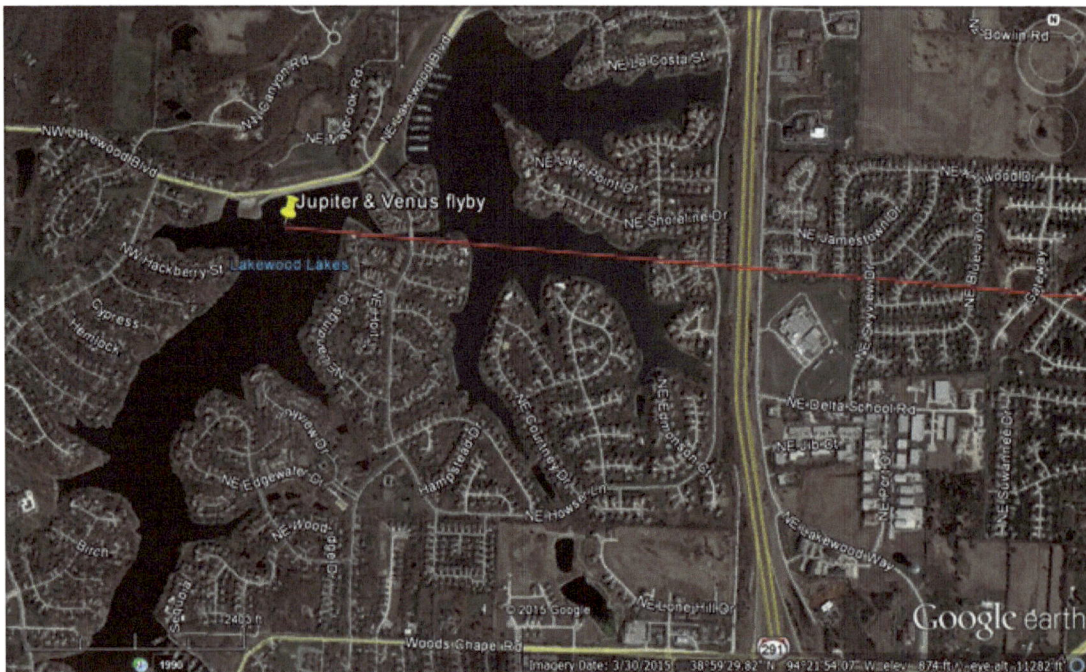

Blue Springs Lake and Lake Jacomo

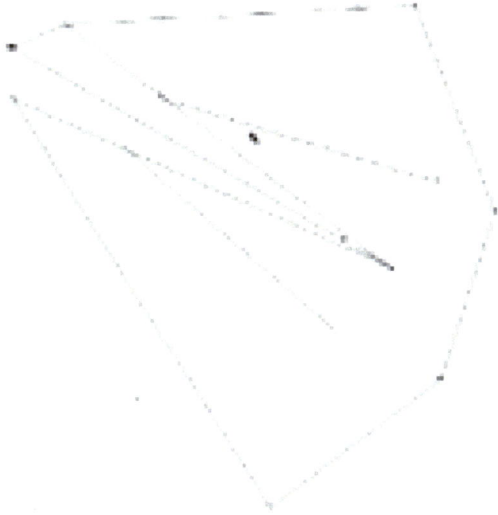

Drawing of Nov. 16, 2014 craft, 117 yards long that is also "football field" sized

were in reality closer, then the speed and size would drop incrementally. Conversely, if they were further away, then the speed and size would be even larger. However, the logic of the sightings being captured over the precise location of the center of the larger of the two lakes is compelling.

At that time of night, Jupiter was positioned over the middle of the Lake, and Venus at the northern end of the Lake from strictly a viewing angle at my perspective. Two of the craft flew between Jupiter and Venus, while the third one was closer to just Venus.

Craft #1 travelled through the screen in 14 frames, or .23 seconds. Craft #2 travelled through the screen in 16 frames, or .27 seconds. Craft #3 took 27 frames due to its extra movements, in .45 seconds.

After calculating the speed of the first craft, it appears to have been travelling at a rate of 821 feet per second, or 560 mph. Of course all of this is based on the estimate that the craft were in the vicinity of Lakewood Lakes. If they

I am amazed at this new technique. Below is an image that shows I was able to pick up enough detail to locate three of Jupiter's four visible moons. With normal telescopes one can see Io, Ganymede, Europa, and Callisto. I had to really push the editing in order to get better clarity for the faint moon images, but they are precisely where the star charts said that the moons were-- at that precise day and time.

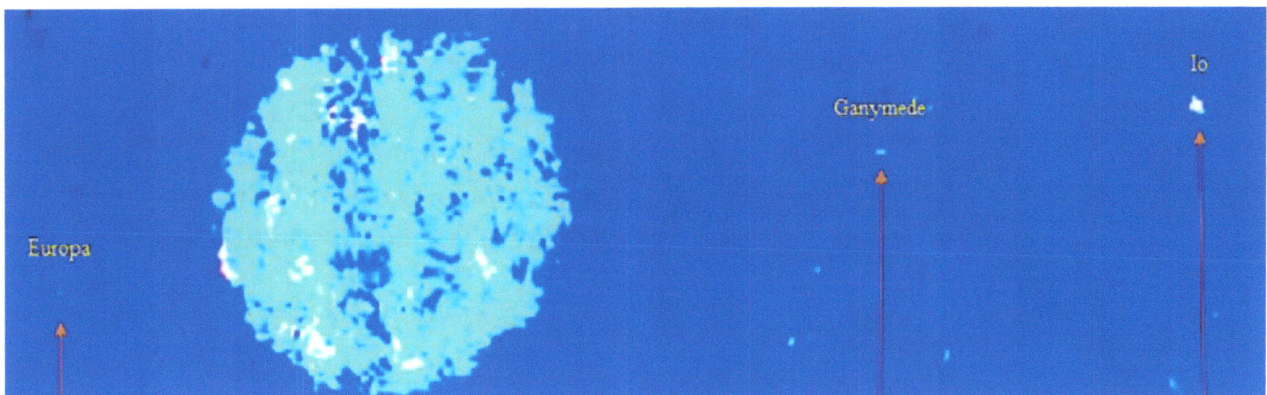

Summary: This may be the first observation and video capture of craft operating in the vicinity of Lakewood Lakes.

Shooting the Sun
MUFON Case # 67423

Monday June 22, 2015 at 7:08 p.m. I was in a parking lot wondering about what the effects would be if I shot my phone camera at the sun to take some video. I had previously thought I had seen something while doing this a few days earlier.

I shot a video that lasted 19 seconds and was aimed directly at the sun which was at 287° or WNW at an altitude of 16°. I noticed an interesting sphere within what appears to be a field.

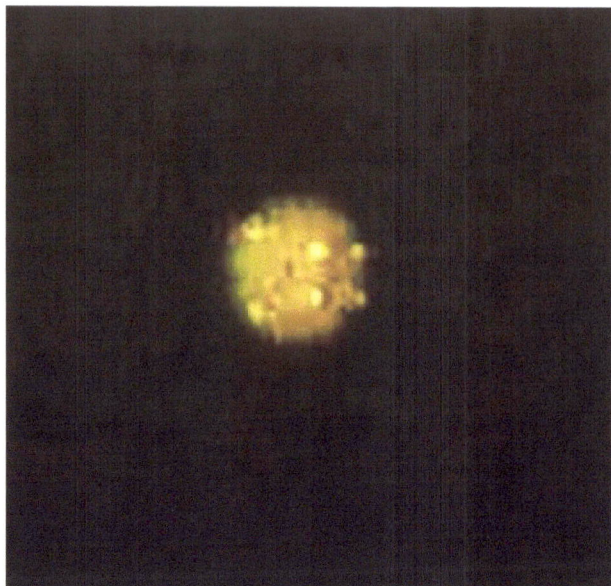

Frame 1:54 at left

Above: Lower section enhanced and close-up

At first I thought that I was seeing was some type of spherical craft, but after further study, I have decided that it was a solar lens flare.

I noticed other slight movements in the video frames, but they are very faint. They also lasted for just one frame. Most were too difficult to get any decent resolution. However, frame 15:54 had an image that was able to be edited to make it viewable.

There was also another frame at 6:30 in which a spot of white appeared in the far right and middle of the image. (See next page)

Frame 15:54

Frame 15:54   enlarged and edited

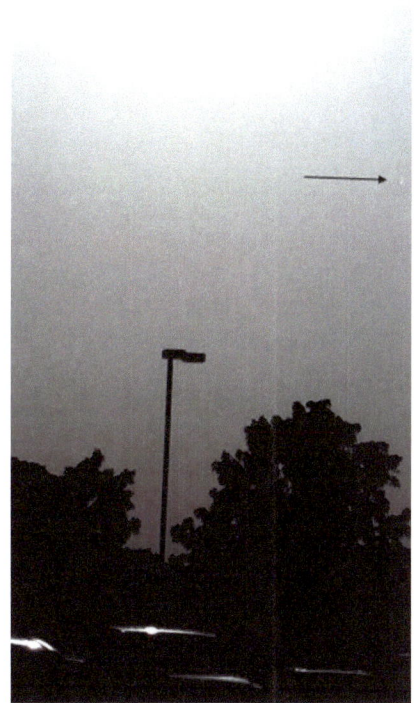

Frame 6:30 White Craft

The light which appeared due to the camera's shuttering down with the sun's brightness allowed more detail to be seen and more clarity in what the nature of the object is in frame 6:30.

Curious about the direction and what lay in the direction of the sun, I checked and discovered that the craft was precisely over the most westward of the two Lakewood Lakes, precisely over the middle of the lake.   (See report about Jupiter-Venus Fly-by).   Two weeks earlier on June 8th, other craft were observed which seem to come from the same general location as well.

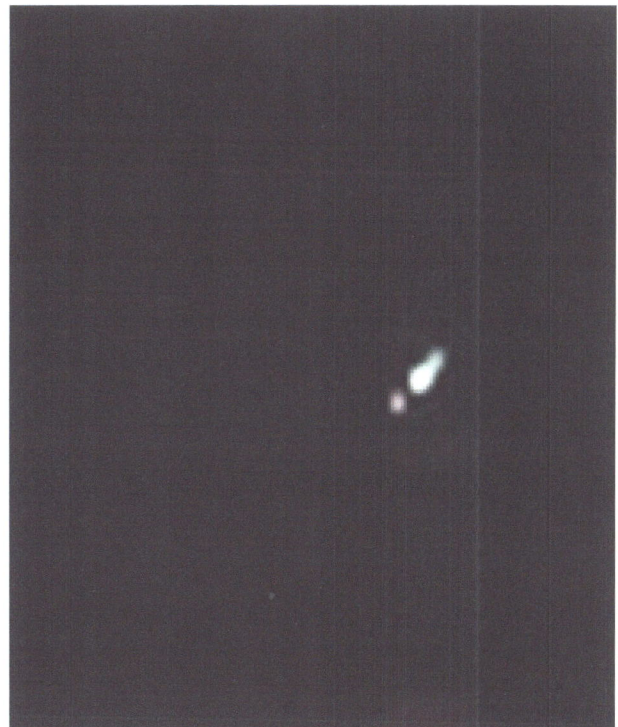

Frame 6:30 edited for clarity

22

The distance this time is 1.25 miles to the center of the lake from my vantage point in the parking lot. Following is the image of the location of the June 8th sighting.

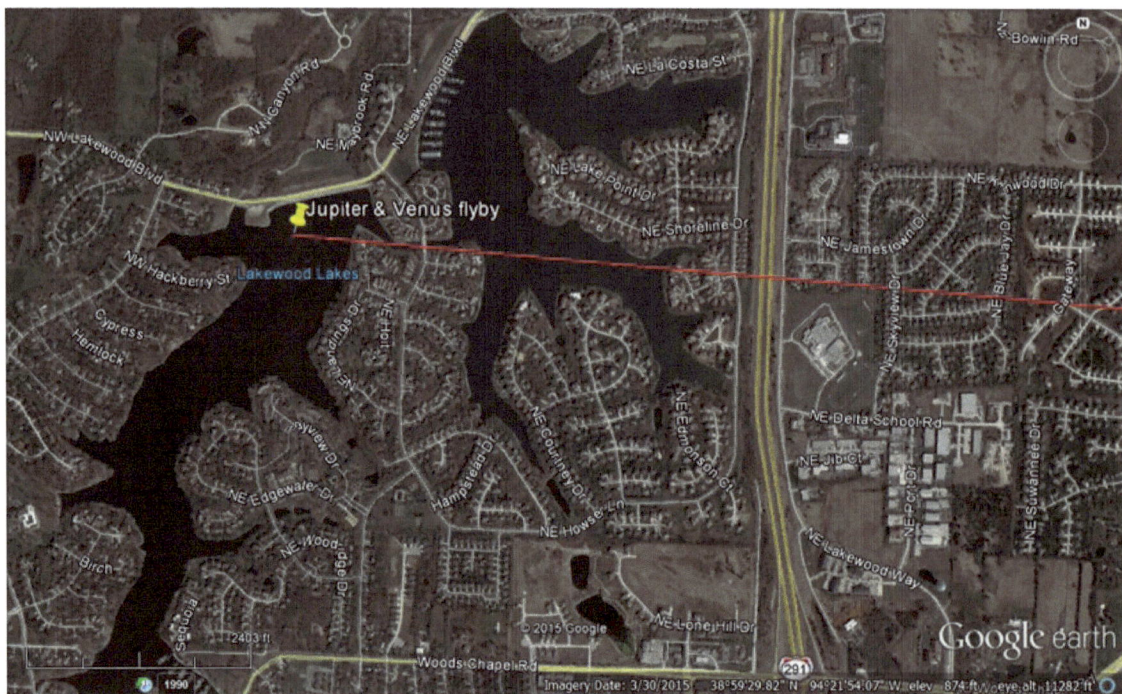

Summary:

Two craft were seen and the fact that they are not in frames before or after speaks to the rapidity of their travel.

The location being identical in the lake is not coincidental since both the Sun and Venus are on the ecliptic plane as well as all planets but Pluto. What is new is the fact that more craft were seen over Lakewood Lake. This makes the 2nd time that Lakewood Lake seems to have craft associated with it.

Chinook Fly-by

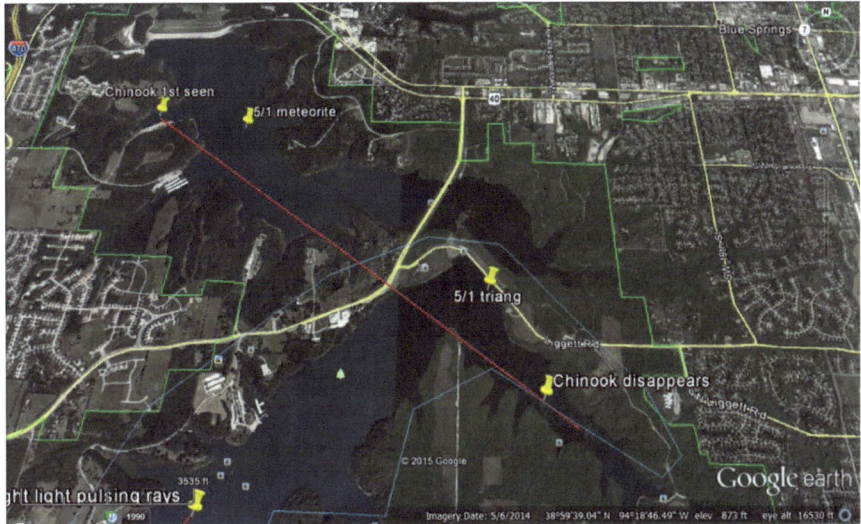

On Wednesday Feb. 11, 2015, in the mid-morning, a military type Chinook helicopter passed over Blue Springs Lake from north to south...passing over Lake Jacomo, and then continued on in the distance.

Sunday July 12, 2015 at 11:02 a.m. a second unmarked military-colored Chinook helicopter flew past the house heading over Blue Springs Lake. This was the second time that day, but I had failed to see it the first time. I noticed it in the distance, approximately .30 miles away, and snapped a couple photos, the first of which was out of focus. This is the view from the back of the house looking toward the lake:

Object 1

Object 2

observed in the above images, two objects happen to be in the air above the Chinook.

The chinook's heading was roughly towards NNW but it had crossed over the western portion of the lake. This is a similar path that occurred in a February 12, 2015 flyby. As can be

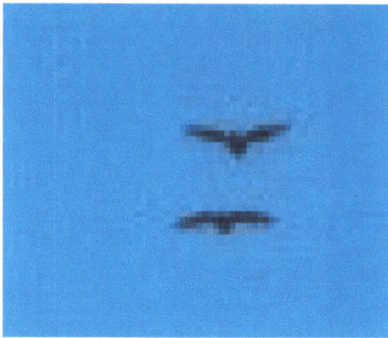

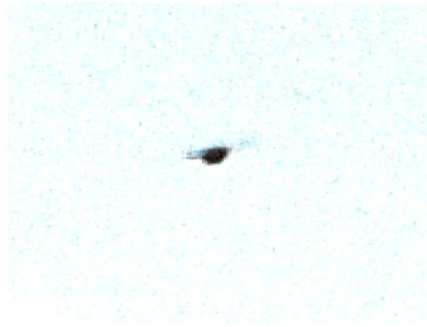

Comparison w/objects   Object # 1   Object # 2

Above, I have introduced an image of some birds that I took several months ago...because sometimes I am asked if the objects that I catch in images are birds. The image on the screen is about the same size as the object #1. One can see bird-like features readily at far left. The two objects #1 and #2 are definitely not birds.

The nature of the two objects is uncertain, but they appear to be some sort of craft. The next image that I took of the Chinook was 6 seconds later but the two craft are nowhere to be seen in that image. The two objects were not visible to the naked eye, only in the photo.

The Chinook **returned** at 11:15 a.m. and I obtained the following image as it flew about 100 yards from the house in a NE heading on the front of the house:

7/18/15 Approx.: 100 yards distant

Return, 13 minutes later

The same day at 4:16 p.m., the same, or a similar Chinook helicopter passed over the same area of both lakes. It flew an identical path to the previous day. Then shortly after this visit, it came by within perhaps 150 - 200 yards of my house... circled, and then made a departing pass through the neighborhood, and then followed Woods Chapel Rd in a westerly direction. It makes one think that the helicopter was looking for something.

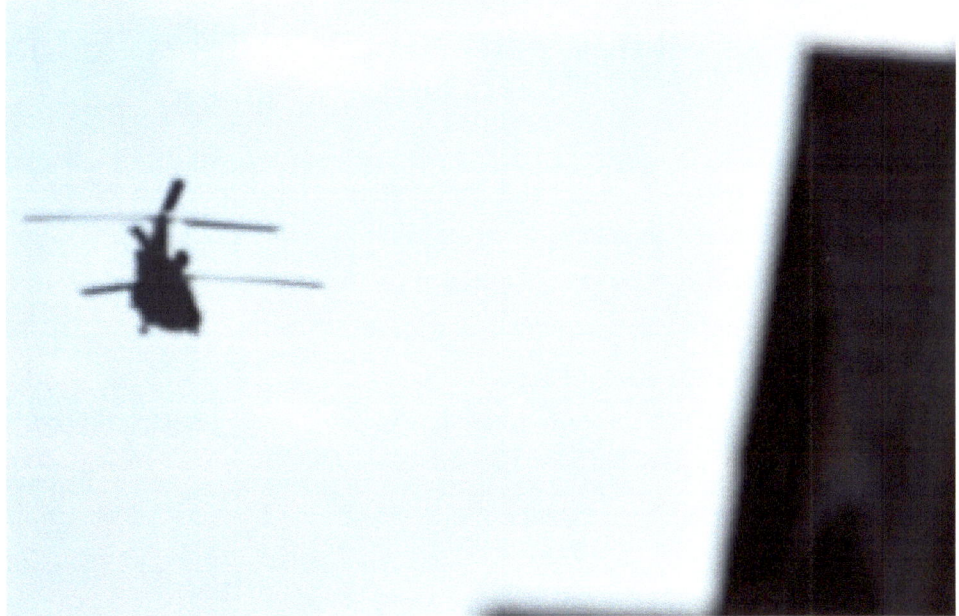

Helicopter Approx.: 800 yards

good camera at the right time, it may likely be that they have their own agendas.

Six days later, on July 18, 2015 at 3:31 p.m., another flyby occurred with a Chinook as it travelled in front of the house again, about 100 yards away. Along with Feb. 11, and 12, April 2, May 4 and now July 12, and 18, the house has been flown by at least six times. It must have been pure luck that I was able to catch the two craft, objects 1 and 2, as the helicopter was travelling.

Conclusion:

Two craft were in the air on July12, 2015. Whether they were being followed by the helicopter, whether they were observing the Chinook, or whether they had their own separate agendas are unknown. Since craft and objects of these types are quite common over the lake, albeit very difficult to see unless caught with a

One White Craft, 3 Dark Spheres, & Many Orbs
MUFON Case # 77915

Saturday, September 19, 2015 I had been noticing some momentary objects in just fractions of a second over Blue Springs Lake. I decided to take a series of four videos beginning at 12:30 pm DST. In the fourth one which lasted 12 seconds, I caught a light grey sphere moving. On a whim, I decided to take a couple photos from the front of the house over a tower in our area.

However, about an hour later I saw two white craft outside of our bedroom window. I ran to the living room and grabbed the camera and recorded a fifth video. A lot of activity was going on I was to discover. This fifth video lasted 51 seconds. The lower of the two craft turned out to be a small Piper type of plane. But what was following it from above was completely something else!! The sequence of events was:

At 12:30 pm DST I took four random videos. Videos 1-3 had nothing of interest. But video 4 did.

At 12:36 pm DST I took a photo of the tower out in front of the house out of curiosity.

At 13:29 pm DST I saw 2 white craft out back, and took the fifth video.

At 13:32 pm DST I wondered if the craft were still visible out front of the house, and took a photo of the plane.

Two White Objects (The 5th video):

Shot at 13:29 DST and toward the NNW (346°) this video caught two white objects in the air at about 30° altitude. The lower one is an airplane. At left is an image of the two craft initially. Note that to the left and upper frame 43:14 there is a small dark sphere above the uppermost white craft. In the next image in frame 46:27, there is also a dark sphere below it. The dark spheres are Fast Movers.

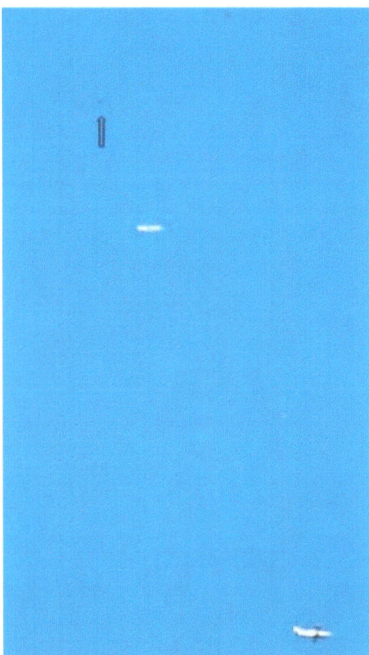

Frame 43:14   (both white craft)

Frame 46:27

Photo of plane

The following still photo taken from the front of our house at 13:32 DST roughly at 90° is of the same plane, but with a dark sphere object in the photo that may have been following it:

Plane   Dark sphere

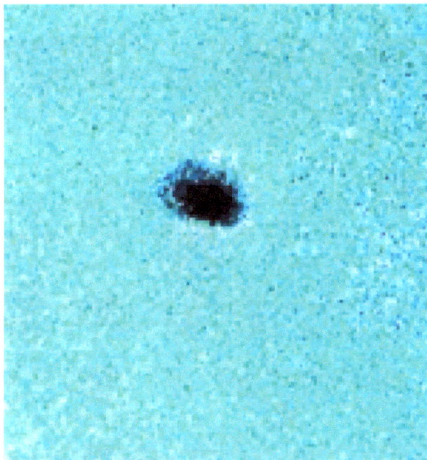

Detail of sphere

Two enhanced images entitled "Frame 36:45" and "Frame 43:26" show more of the detail of the white craft that was following the airplane. At times the craft seemed to lengthen. At other times it was not in two pieces but a single white craft. Since a Piper type aircraft is 22 feet long, and the craft following the plane was 60% of it's size (after measurement), this makes the craft 13 feet in length.

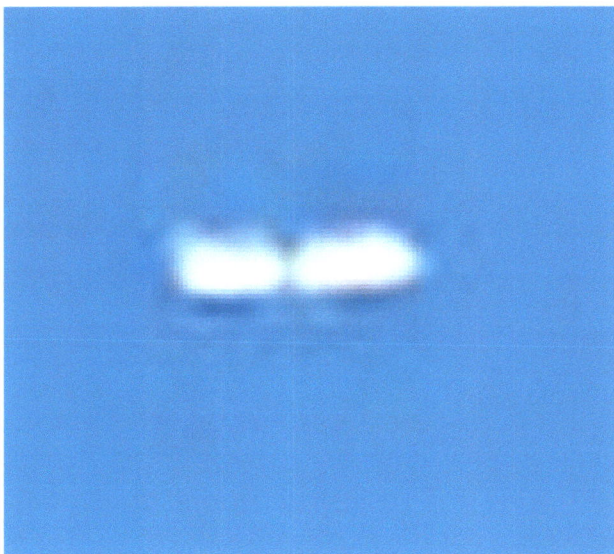

Frame 36:45

Frame 43:26

Photo of tower:

At 12:36 DST the following image of the tower was taken at 96° (east), and after working to enhance the photo, and inverting the colors, shows at least 18 dark orb-like craft around it:

From the 4th video:

At 12:33 DST, the following video shot at 346° (NNW) caught some dark spheres in frames :00, :06, and frame :19. This photomerge of the three frames shows how the Fast Mover traveled in 1/3 of a second making an angular direction change of 50° in 13/60 of a second.

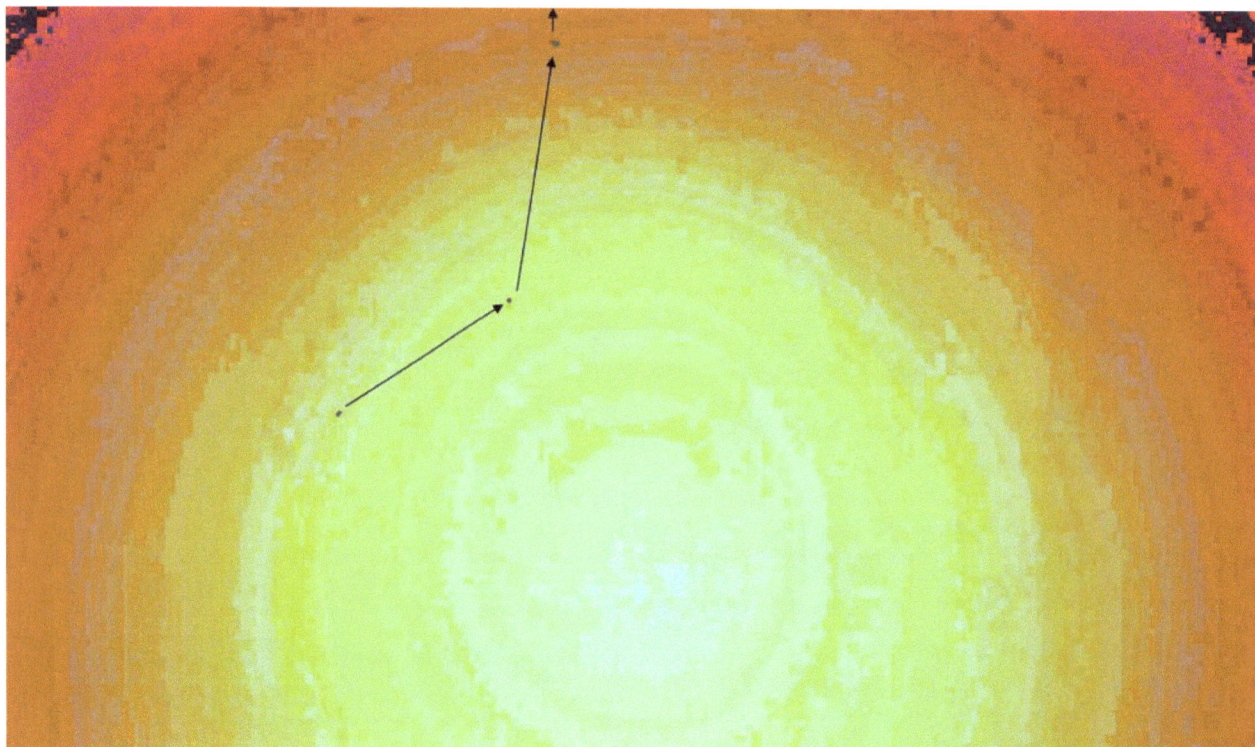

The following image of the craft was the clearest in frame :19 and has been cropped and enhanced...

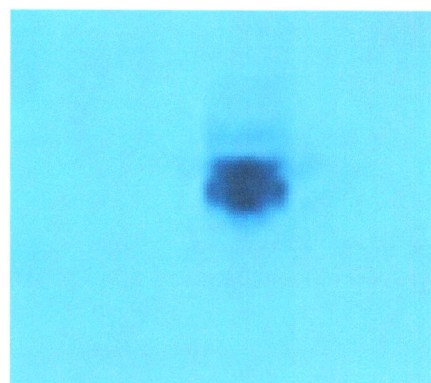

Summary:

At least one dark craft was seen during the shooting of the 4th video at 12:33 DST in the NNW. It was seen travelling very rapidly through the viewfinder in 1/3 of a second, making an abrupt directional change of 50° straight toward the zenith.

Three minutes later, at 12:36 DST at 90° (east), at least 18 different dark orbs are seen surrounding the tower.

An hour later at 13:29 DST, at 346° (NNW), the 5th video of the two white craft shows that one was an airplane, and the other is an unknown UAV which seemed to be stalking it, that was 13 feet long.

In addition to the white craft, two small dark spheres can be seen surrounding the white craft.

A total of at least 22 craft were seen over the span of an hour in this report.

Case 76946 (Full Spectrum Camera Images) was shot on the same day, about an hour after the last image in this report. Four craft were seen in that report. Thus in the span of two hours that day, 26 fast moving craft were filmed and reported.

Full Spectrum Camera Images of Fast Movers

MUFON Case # 76946

Full frame 4:10

In August 2015, I purchased a full spectrum adapted video camera. I had not had a chance to use it. So on Saturday September 19, 2015 at 2:34 pm, I took some random video. After viewing the camera screen quickly, I forgot about it. When I happened to look more closely at it on my computer in frame by frame analysis at the video on February 20, 2016, I noticed fast movers.

There seem to be two kinds of craft in the 30.3 seconds video. The larger one which initially seemed like a walnut appeared three places briefly in the video. It was not a fast mover. The smaller ones resembled black dots, and were definitely fast movers!

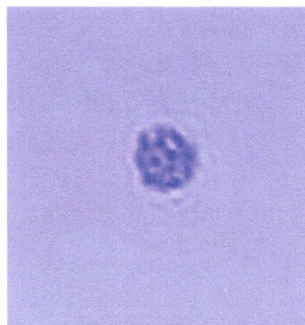

Frame 4:10

Frame 4:44

The Larger Craft

The larger craft began to appear in frame 2:20 and vanished by frame 2:27 moving to the right in .11 seconds.

It reappeared at frame 4:00 moving to the right, reversing and going left, danced upward and left and vanished at frame 5:08, a total of 1.13 seconds.

Finally it reappeared from frame 26:04 until

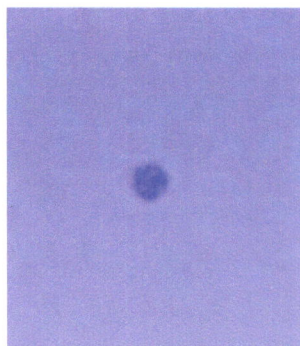

Frame 26:16

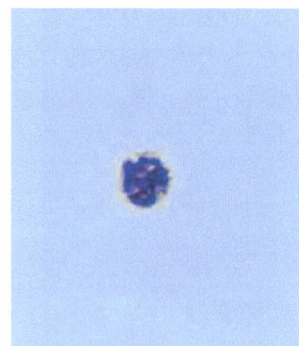

Frame 26:17

frame 27:24, lasting 1.33 seconds.

The Smaller Craft

Dot shaped craft shot up and to the left at 7:52 in 11 frames at a 51° angle in .18 seconds.

A second dot shaped craft shot down and to the right at a 31° angle in frames 24:29 through 24:47 in .36 seconds.

Then a third dot shaped craft shot down and to the right at a 45° angle in frames 28:42 to 28:44 in 3/60th of a second.

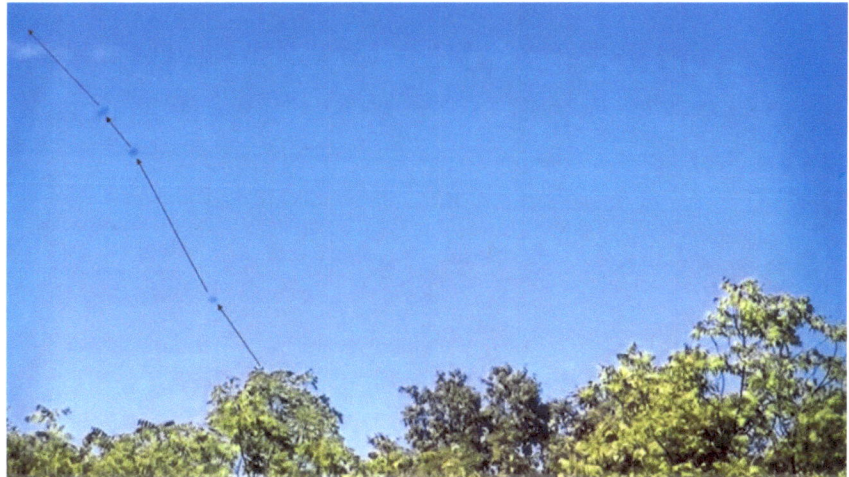

1st dot shape craft--Photomerge of frames 7:53, 7:57, and 7:59 and trajectory

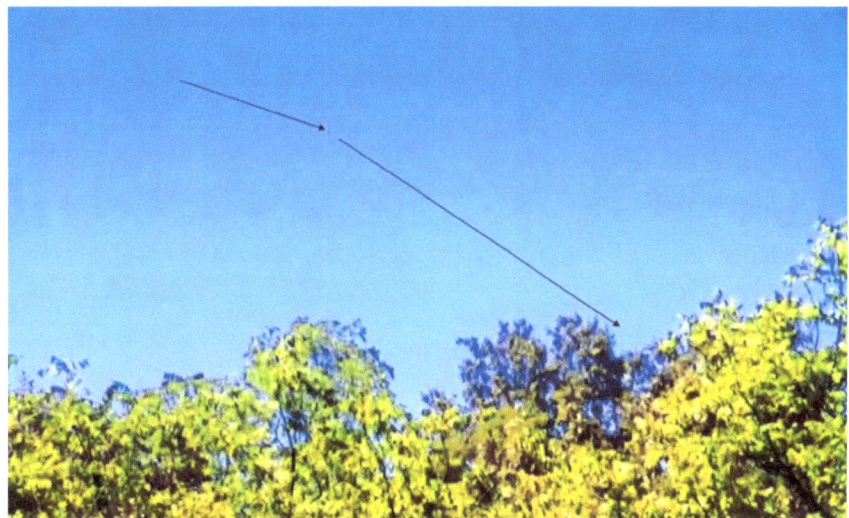

Below: 3rd dot shape craft Photomerge of frames 28:42 to 28:44 and trajectory:

2nd dot shape craft Photomerge of frames 24:35 and 24:40 and trajectory

small dot shape craft Frame 28:43

Analysis:

Whether or not the infrared and UV light aided in identifying these craft will be interesting to study in the future. Many other reports have been filed by this investigator of similar types of craft coming from and going to Blue Springs Lake. The upward craft moving to the left seems to have come from nearly NNW at 336° in the lake. The first downward to the right

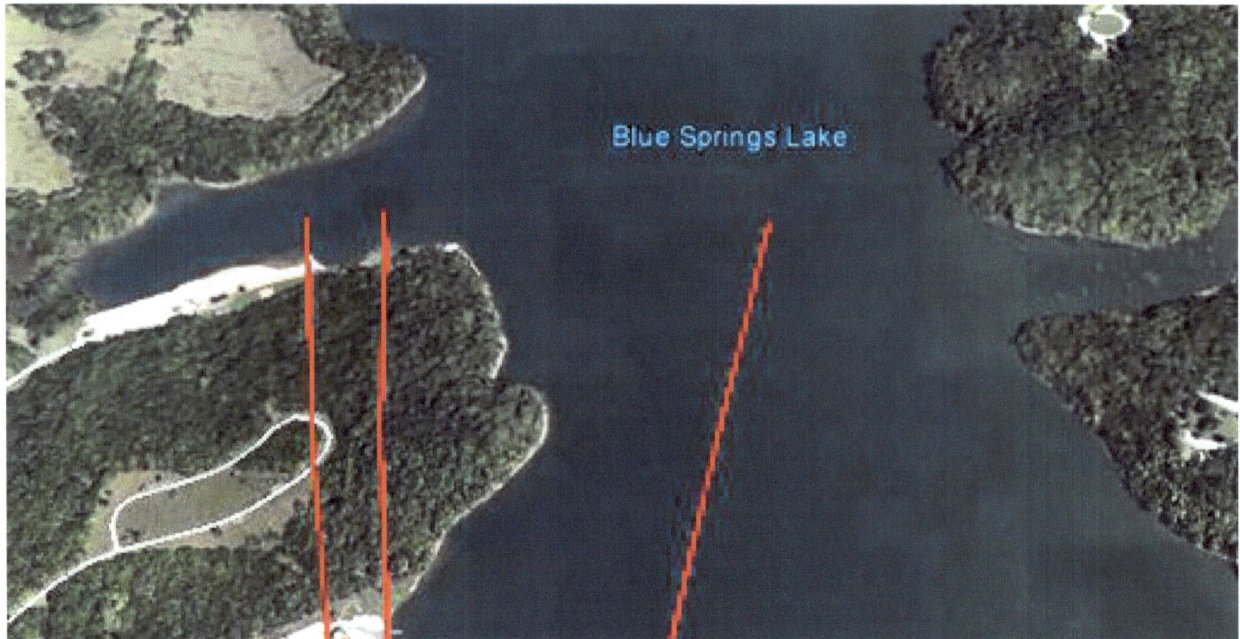

Case # 60295 craft emerged here

moving craft dove into the area of the lake at 0°, or due North. It is assumed that this is the same location in the lake where other craft have been seen repeatedly-- which is roughly 1 mile away.

The size of the **larger** craft appear to be approximately 15ft in diameter based on the size of previous similar craft.

The distance is estimated at 1 mile as well. It seems to have remained at approximately 12° elevation most of the time, moving back and forth, going in and out of view.

The size of the dot shaped smaller craft appear to be about 2.8 feet, being nearly 19% of the size of the larger one. It is interesting to note that the path taken into the lake by the **third** craft which went downward appears to go precisely to the same location into the lake at 0° north where the **upward** craft came from and the **second downward** craft also went.

Conclusion:

The behavior of the walnut-shaped larger sphere was odd in that it would go from invisibility, to visible, move back and forth...and then simply vanished again near the end of the video. It hovered and seemed to morph to slightly different shapes. The smaller dot-shaped craft travelled at between 275 mph and 330 mph.

The fact that at least two of the smaller downward craft entered at the exact same location cannot be coincidence. Case # 60295 reported an object coming from the lake and doing a zig zag in the air...coming from almost precisely the same location as the downward craft #1 entered the lake. The odds of this is also 332,212 to 1. The dual coincidences taken together actually increase the odds significantly to

110,364.81 $x10^6$. Clearly, these are common locations used by craft.

Mass UFO Sighting in Kansas City June 20, 2019

Editor's note: Thousands of people witnessed UFOs over greater Kansas City on June 20, 2019 for over six hours. During that time, these white objects remained stationary at 2,900 feet. DARPA released a statement that afternoon with no mention of what was going on in K.C., but apparently in reaction to the news stories. The statement mentioned that several high-altitude weather balloons had been released three days prior on the East coast of the U.S. These balloons are designed to be located above 75,000 feet in altitude and move with the jet stream, which goes from west to east. After the release by DARPA, most news outlets assumed that the objects were weather balloons. Missouri MUFON is still investigating these sightings and the case remains open. Visit www.missourimufon.com for more information about this event.

What the media did not cover is the fact that after careful analysis of four of the videos submitted by witnesses, Wayne Lawrence was able to ascertain that there were no less than 20 additional fast-moving objects around the thee larger objects. The following three cases are from this event.

Two Motionless White Objects

Liberty, MO
MUFON Case # 101408

On Thursday June 20, 2019 at 7:39 pm in Liberty, MO, a witness reported seeing two motionless white objects in the sky. They were directly north at an elevation of 17 degrees. The witness took seven still photos and five videos over a period from 7:39 pm to 7:58 pm—although he watched the objects for a period of 35 minutes. What he didn't see was Fast Movers.

One of the still photos of the white objects.

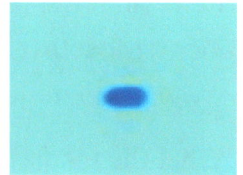

Video 2: Black craft flying under one of the white objects at 8,217 mph.

Close-up

Video 3: Black craft very rapidly flying underneath an airplane that was headed toward the white objects.

Frame 5:2` shows an object the flew through one frame underneath the airplane

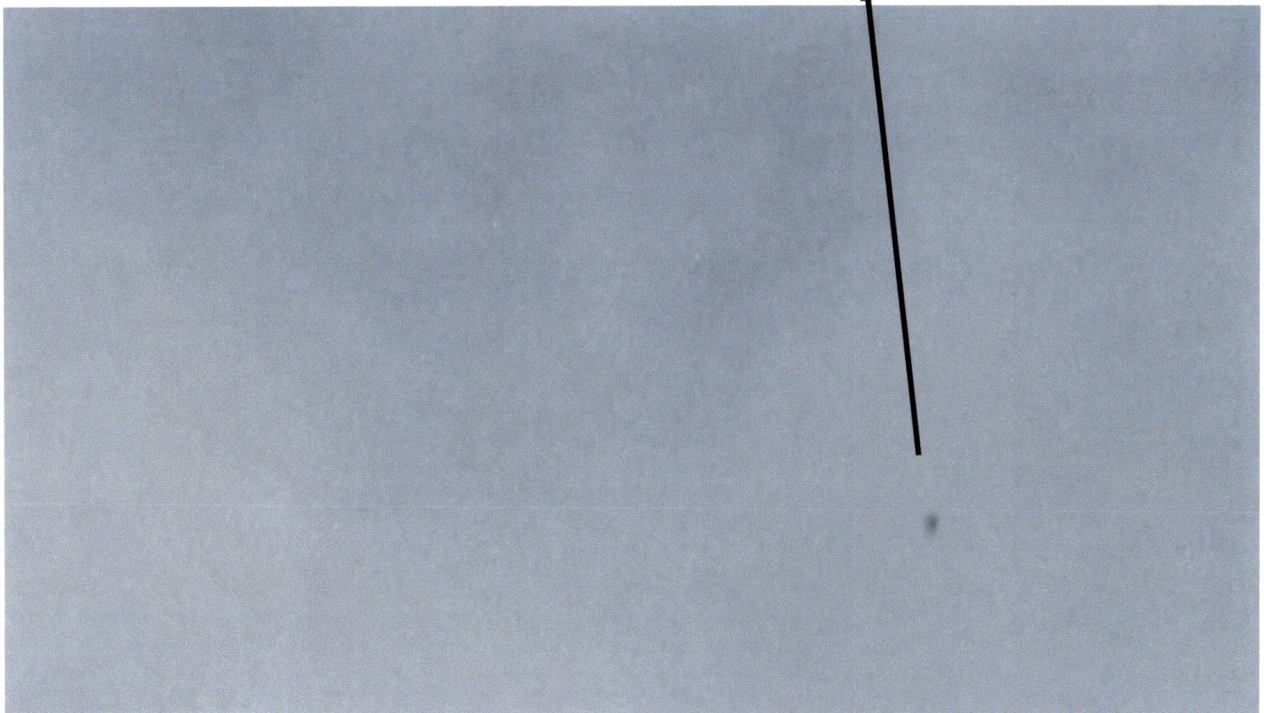

In frame 31:10 of video 5, a grey blob exits in two frames

In most of the videos (5 of six), one can see black craft making flyby passes at both white objects. These are travelling very fast and cannot be seen by naked eyes when playing the video. Only by progressing frame by frame can one catch them. One of the black craft was flat and disk shaped. **Its speed calculated to 8,217 mph.** Most of the others were small black dots or spheres.

In summary, at least 6 other black Fast Movers were seen making a total of over 20 flyby passes. It is also possible that there were 20+ craft. We just see them exiting various sides of the screen.

The two white craft were seen all over the Kansas City area by thousands who either saw it with their own eyes, or saw the television reports of them by at least two stations. However, until the frame by frame analysis, no one knew about the Fast Movers.

Two Lights seen from Blue Springs, Missouri
Thursday June 20, 2019
MUFON Case # 101330

At 7:50 pm, during the same time that the witness in Liberty, MO was also seeing the two white objects, a second witness in Blue Springs, MO took four videos of the objects. In three of the videos, black Fast Movers were caught on film

A frame from video 1.

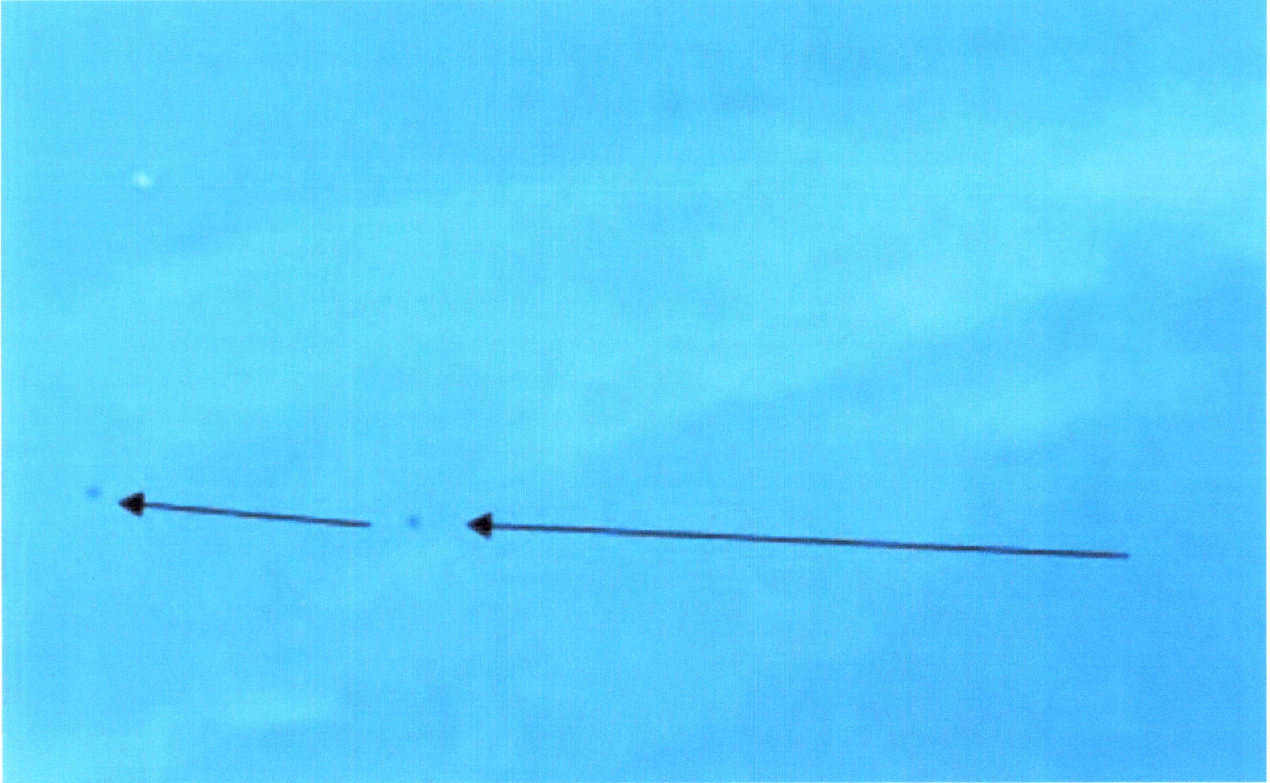

In frames 14:13 through 14:20 of the first video, a Fast Moving black craft entered from the right and flew under the white object

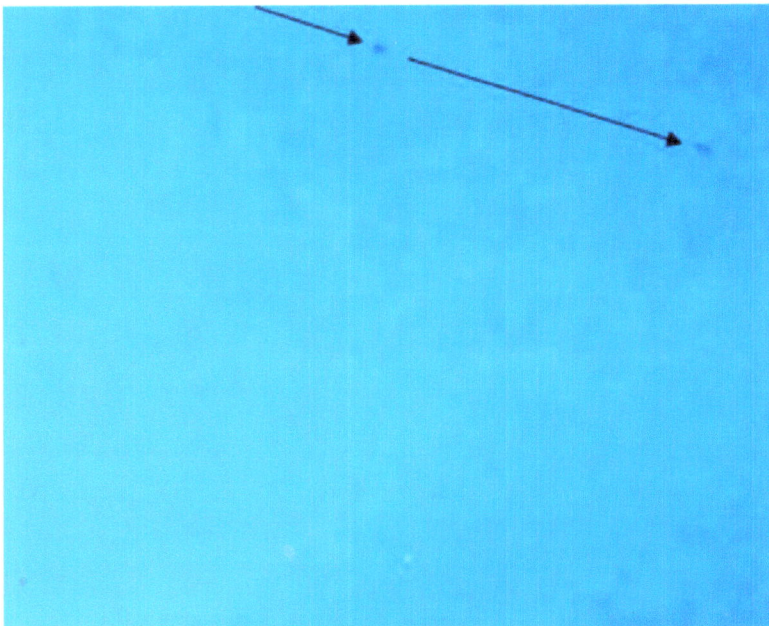

In frames 33:14-15 of video # 3 a black craft flew through.

There were also a grey sphere and two small black spheres that were seen in video #2 not pictured here. The Fast Movers viewed from Blue Springs were different from the ones seen in Liberty.

Two White Objects—Berkley Riverfront Park, KC, MO

Thursday June 20, 2019

MUFON Case # 101349

At 2:00 pm on the same day that the sightings in Liberty and Blue Springs, MO were seen—the same two white craft were seen from downtown Kansas City at a local park. However this was nearly six hours earlier! The witness took some photos and a video lasting about a minute.

The white spot near the middle of this image is one of the two white craft. The second was too hard to see or photograph and out of the range of this image.

At 24 seconds into the video, two black Fast Moving craft approached the left white object. One vanished and the other continued on past the white object.

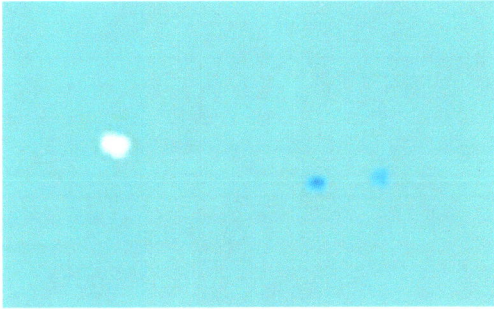

2 black craft approaching the white object.

At 50 seconds, another black craft began at the lower right side of the screen and travelled to the left under the white object:

Black sphere nearly underneath

Black sphere entering from the right.

At 54 seconds, 4 white craft appeared. A fifth one streaked downward to the right and out of view. The four additional craft at one point all lined up nearly on top of each other. At other times, two of them formed a triangle, made a fairly straight line, and then made a second triangular shape

Two of the better images are shown below:

Two pairs of spherical craft flew above the white object

Needless to say, all these fast movers were not noticeable by viewing the video. It had to be stopped frame by frame in order to see these objects

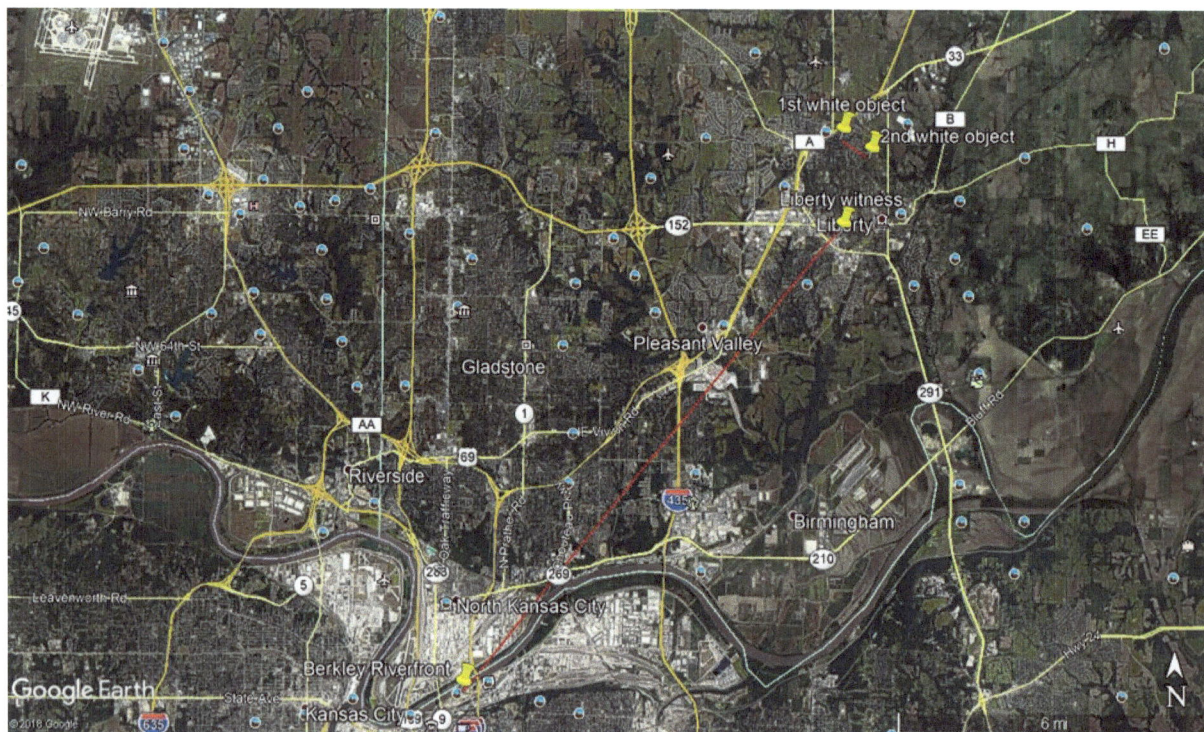

In the map of the area, in the lower left the yellow pin shows the location of the Berkley River-front sighting. A red line goes up to the area of the Liberty sighting.

What are the odds of coincidentally seeing an object from Berkley Riverfront Park in the same direction as the Liberty sighting six hours before they were noticed in Liberty by the witness? **74,228,620 to 1**

DATA from Berkley Riverfront:

Directional heading: 40 degrees (NE)

Elevation from witness: 20 degrees

Altitude: ? Feet (not enough data)

<u>Distance</u> from witness: 2-3 miles?

Actual size of object: 36.4 feet (if same as in Liberty)

Another Berkley Riverfront Sighting
July 2, 2019

At 11:19 am on Tuesday July 2nd, 2019, I visited the same location in which the previous Riverfront witness had her sighting on June 20, 2019. My intention was to ascertain what the witness' angle of observation was 12 days earlier by comparing images from the witness with an actual reading using an inclinometer.

I took two random videos. On both videos I caught black Fast Movers.

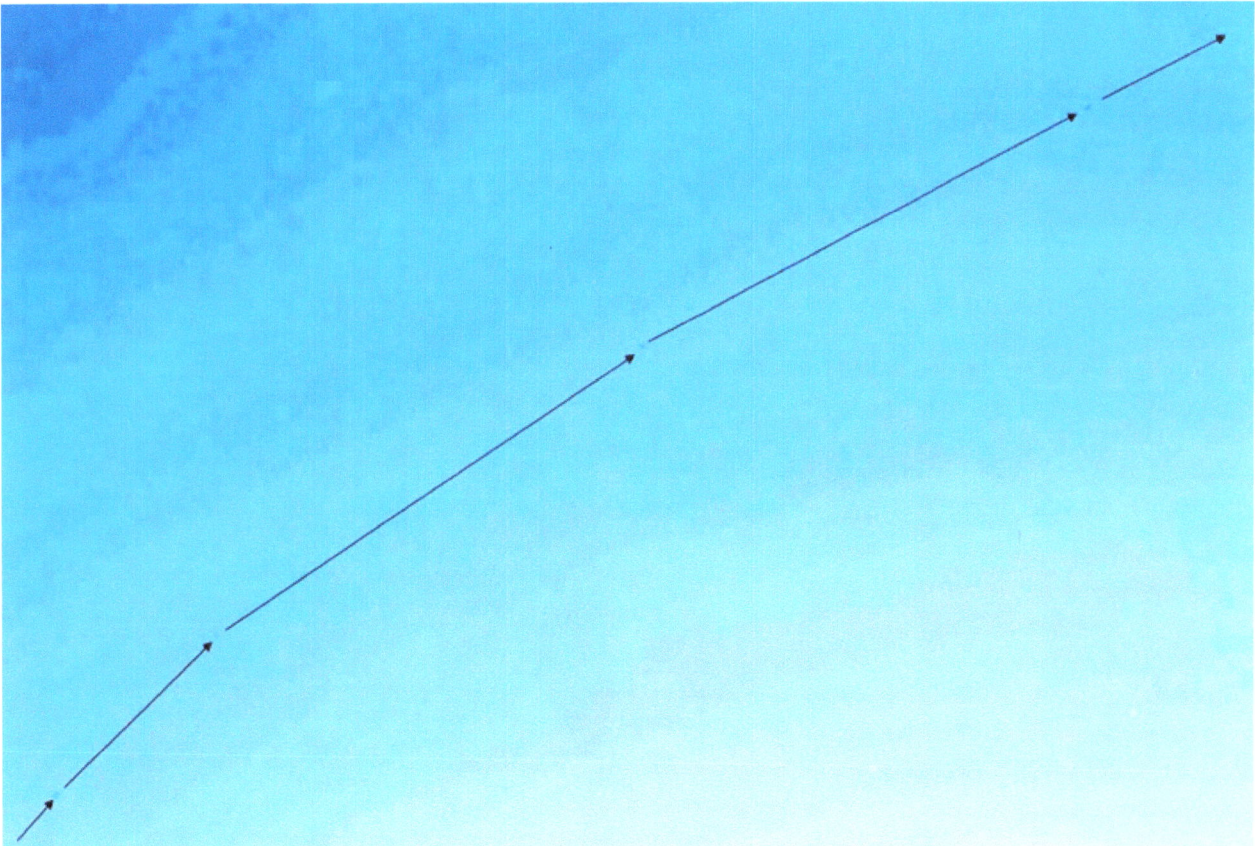

Video #1: Frames 1:17, 1:21, 1:27, 2:01 photomerged.

Seven Craft over Venus

At 6:52 pm on February 3rd, 2017 I was photographing the planet Venus and experimenting with the focus by backing it out and causing the image to expand. In the past this has proven to be useful in seeing some detail which normally are not visible. At the date and time that it was photographed, the planet Venus was at an elevation of 28.7° from the horizon and at 247° azimuth, or WSW.

In degrees, Venus is .63 arcminutes or .0105°. When compared to the size of the image of the disc of the planet, the craft are 83.3 times smaller at .000126°.

The entire satellite database was consulted to see if any were in the physical space of 28.7° and 247° azimuth (WSW) at the date and time of these photos, and none were.

While looking at three of the images, I noticed that there was some type of craft in the images. All told, seven images were taken but only three show craft in them. I just happened to catch the craft as they were crossing the field of view of the image.

There are three colors to these craft. Image #1 has a small black sphere in the southern area of Venus that initially drew the witness to investigate it. On further editing the image the I became aware of the white craft above Venus. There are also craft that are somewhat translucent in images #2 and #3. Both the black craft and the translucent craft appear to have a field of some kind of field surrounding the spheres because it creates a perfectly round "halo"

effect surrounding them.

In image #1, the dark spot near 7 o'clock is enhanced in the four Image 1 edits. Also, one can see a white craft off of the face of Venus near 1 o'clock.

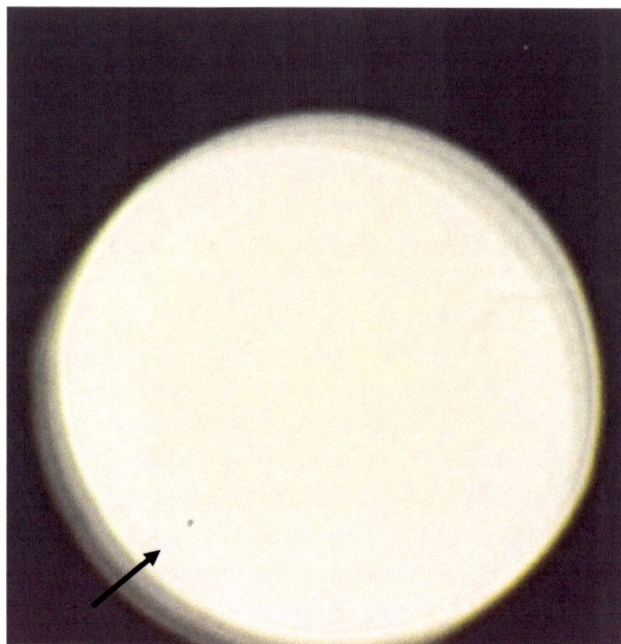

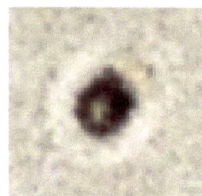 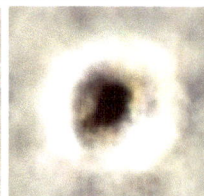 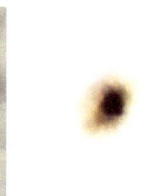

First edit   Second edit   Final edit

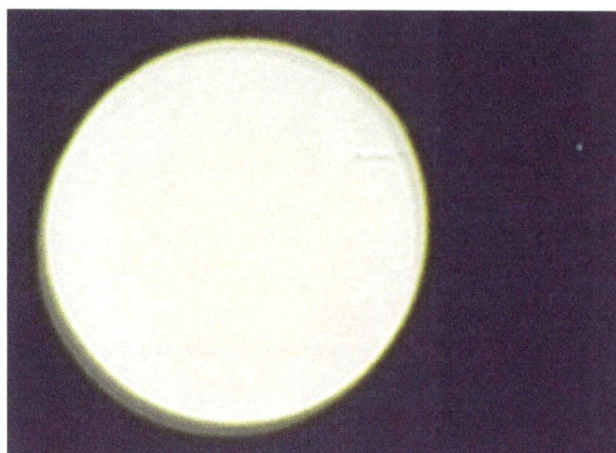

Image #2

43

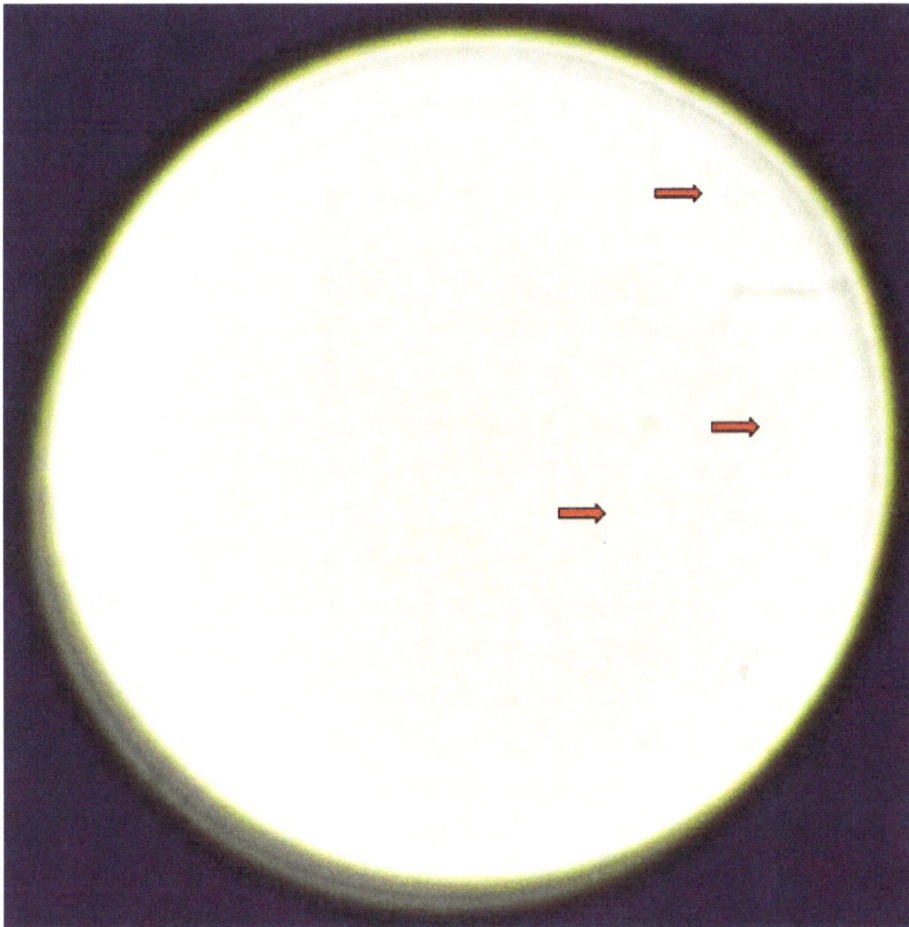

Image #2 was taken 7 seconds after #1. There appear to be three other craft inside the rim of the planet. One is near 1 o'clock, another is not too far from the center of the image, and yet another is at 3 o'clock.

Image #3 below was taken 31 seconds later from Image #2. The object is located at 3 o'clock. This craft is 4 times larger (or closer to the camera) than the three in the previous image.

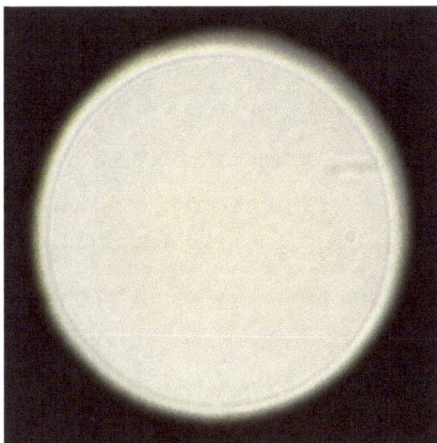

Image #3

and two enhancements at left

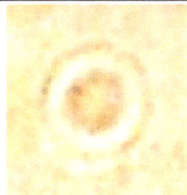

The only reason that I saw craft in these photos is because I blew up the image using the change in focus, and then further enlarged the images on the computer. I saw nothing particularly interesting as I snapped the photos.

We cannot know for certain how high these objects were and if they were still in Earth's atmosphere or outside it. There is no reference information for it. All that we can state is that image #2 was taken 7 seconds after the first image, and image #3 was 31 seconds after #2. It does not appear that any of the seven craft were in the same photo twice. Therefore it is reasonable to assume that these and maybe other craft were traveling rather fast, as they tend to do at high altitudes. It would have been interesting to take yet other photos, even video, to see if there were other craft traveling through.

Blue Cone in the Sky

On June 14, 2017 at about 8:00 p.m., a witness saw a strange cone of blue light in the sky while driving home. She believed it to be emanating somewhere from either the Holden, MO or the Chilhowee, MO general area.

#1 craft—enhanced

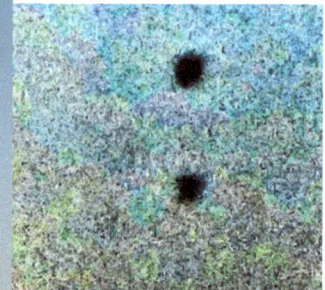

2 craft—enhanced

First photo with 1 craft   Second photo with 2 craft

What the nature of the craft are in both images remains to be discovered. There is no way to know how long the craft were in the air, their destination, speed or size for certain. We do know that the craft in the 1st image was not in the same location by the time that 2 minutes had elapsed and the 2nd photo was taken.

The fact that they all appear to be associated with the blue light is interesting. Did they originate from the lower part of the blue cone near the ground and were being released?

With such few other facts, one can only conjecture. The spherical craft could have been coincidentally in the Lee's Summit area where the photos were taken. Yet the photo taken two miles later by Todd George Rd still has craft associated with the blue light. One can hypothesize that the craft have a relationship with the blue light. The witness believed that the light (and perhaps the craft) were about 43 miles away (in the Chilhowee area) when the photographs were being taken, then the first sphere would calculate to be approximately 4.12 miles

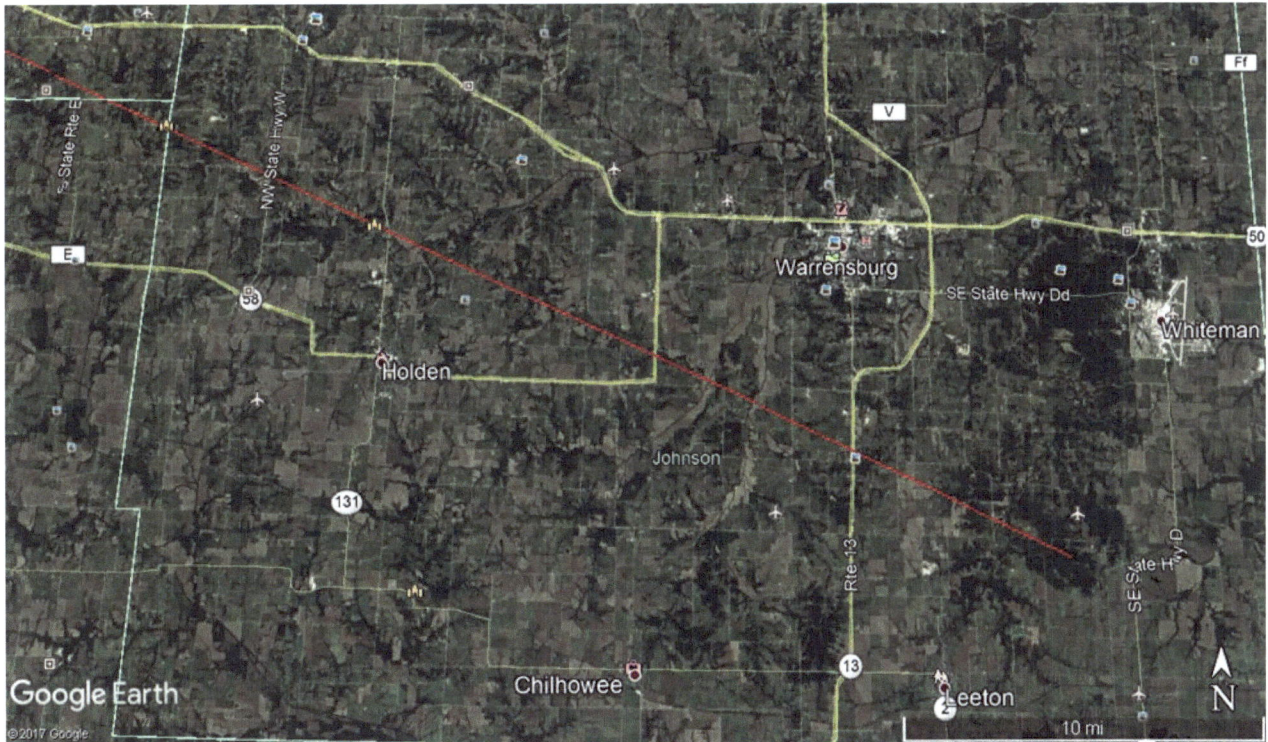

The red line at far left is the direction from which the witness viewed the blue light. The red line ends in an unpopulated rural forested area approximately 6 miles from Whiteman AFB.

in altitude, and 64 feet in diameter.

After traveling in her car for two minutes, the craft do not appear to have changed much in size. Most likely they were closer to the source of the blue cone of light. The fact is that in each photo taken by the witness, we find craft. That too is interesting.

Glowing Sphere
MUFON Case # 105862

On Tuesday, January 21, 2020 in the northern part of the Kansas City, MO area, a witness noticed a bright glowing Sphere hovering in mid-air. It appeared to be bobbing up and down and swaying left and right, and at other times just hovering. The time was 6:10 pm and he proceeded to take two videos each lasting nearly 5 minutes in length. Here is an image of the sphere near the top of the image. One can see three other craft near the bottom that are in front of a tree and much closer:

The same object has also reappeared on January 25, January 31, Feb. 1 and February 6 to the same location in the sky. Most of the videos and selected frames from those videos are of a craft that is virtually motionless, hovering in place. However in one of the videos from January 25, there appear other bright smaller objects in all directions surrounding the main Sphere. The following are a selection of the images:

Frame 20:28 at 8 O'clock

Frame 29:23 at 3 O'clock

Frame 29:29 at 7 O'clock

Frame 51:09 at 9 O'clock

Frame 1:10:00 at 4 O'clock

Frame 1:34:04 at 11 O'clock

The characteristic of these "extra craft" that is so unusual is that they appear only for 1 frame, and then are not seen again for between 2 seconds and 45 seconds when they reappear in the same area. One possibility is that they are travelling so rapidly that they are only caught in one frame before they exit the screen. The normal progress of a "fast mover" moving across the screen in several frames is not seen in this case.

Another possibility is that these smaller craft are just cloaking and uncloaking themselves. Yet another potential is that they are appearing from another dimension just briefly. At this point there is too little information to be able to assess what is actually going on.

The poor video quality is such that there seem to be other "craft" that are not as distinct and bright as the ones above. It does seem that there are multiple other ones doing the same thing. They were too faint and had too little definition to be able to be captured and viewed in frames. There were dozens if not hundreds of those.

It is not known if they are related to the same Bright Sphere, or if they were observing the sphere out of curiosity. I have noticed a similar phenomenon in many of the fast movers reports because they do seem to be interested in visiting the major craft that are in the sky. Could they be another extraterrestrial's version of a drone? At this point, that is just a hypothesis.

Disc-shaped Craft

MUFON Case # 77979

On Thursday July 21, 2016 I took a random video that lasted 39 seconds at 15:28 DST. During the video a flat or possibly disc-shaped craft travelled from frame 18:35 to 19:05, or ½ of a second. The angle of rapid descent was at 14° from the right side of the frame to the left. Nothing else was seen in the remainder of the video. The angle of elevation when it entered the screen was 12°, although it clearly came from much higher.

The direction of initial view was 350° or nearly North, and it disappeared from the camera view at 336° or NNW-- although it continued in a straight line ending at about 320° or NW where it likely ended its flight. It disappeared on the west side of Blue Springs Lake, interestingly, into another much smaller lake.

In the image above, one sees the direction of the flight of the craft from frames 18:35 to 19:05.

From right to left, location of entry, departure from view in camera, and the final descent location.

Craft from frame 18:47

Enhanced view of frame 18:47

End location craft in small unnamed lake

Analysis:

In 16 frames, the craft covered a distance of .3 miles in ½ of a second, which converts to **2,160** mph. Most likely it entered the small body of water, although not actually seen doing so. The direction of the 14° descent, by projection, ended NW at 320°. This small lake is about 1600 feet west of Blue Springs Lake. The size of the object appears to be close to 35 feet when calculated. This Fast Mover was only seen initially with frame by frame analysis.

CONCLUSION:

The video evidence shows a craft flying at an incredible speed of 2,160 mph fractions of a second before ending it's travel within 3/10 of a mile demonstrates a control unknown to our engineering. The physics of such flying make it impossible to do with known technology if it were a human vehicle. This makes it to be of unknown origin, and not human made.

END NOTE

It appears to this investigator that these Fast Movers are more common than one would imagine. This suggests some questions that surround them:

- Just how prevalent are these Fast Movers world-wide?
- Do they tend to stay under lakes with home bases below the lakes?
- Do they follow other craft out of curiosity? Or is it just luck that we catch them?
- What is the nature of these Fast Movers? Are they piloted by sentient beings?
- Are fast movers just versions of an extra-terrestrial drone?
- If extraterrestrial, what star(s) are they from?
- Where are they going when we catch them?

We have examined over 21 reports containing at least 90 craft that were fast moving and unable to be seen by the naked eye. It might be interesting to randomly take videos every hour of the day during daylight to see what is occurring. I guess that could be a future undertaking!

Methods to establish distance, height, size and speed of objects

By Wayne Lawrence

Formula:

$\underline{a°}$ = tangent Θ   -- where theta (Θ) is the angle of observation of the object
a

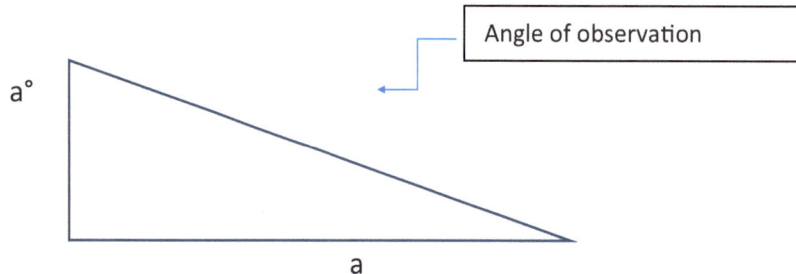

If a person knows any two of the following…

    Either the distance of a°

    The distance of "a"

    Or the angle of observation

…then you can calculate the remaining item using the formula.

The distance of a° can be the height to an object that was observed. It can also be the size of a craft. When the angle of observation is known, for example 20 degrees of elevation, one can then measure the size of the craft on the computer screen in millimeters using a ruler. On the screen, measure the distance from the ground to the object. Then measure the size of the object itself on the screen. Next, divide the size of the object by the height of the object. That tells you what percentage the object is compared to the overall height of the object. Very important information!

HEIGHT, DISTANCE, AND ANGLE

Let us suppose that the size of the object is 2 mm. Next we measure the distance of the object from the ground up to it, for example let's say it was 40 mm on the computer screen. The size of the object is 2mm divided by 40mm = 1/20th, or .05 in decimal. Since the angle of elevation was 20 degrees, multiply it by .05 and you get 1 degree. That is the angular size of the object, or the angle of observation.

Now let's assume that the height of the object is known because it was going in and out of clouds. By finding the temperature and the dewpoint, at www.Wunderground.com , then go to

https://www.easycalculation.com/weather/cloud-base.php and plug in the data for temperature and dewpoint, and you will get the cloud base height. That is the height of the object a°.

So, using our example, we can solve for the distance "a" in the following formula:

$\underline{a°}$ = tangent Θ
a

Let's assume that the temperature was 67 degrees, and the dewpoint was 50 degrees. The resulting cloud base is 3,863 feet. So " a° " is 3,863 feet.
We know the angle of observation to be 20 degrees.

So the formula is: $\underline{3,863}$ = tangent of 20 degrees
                                a
You can look up online for the tangent of 20 degrees. It is .3639.

The formula now is: $\underline{3,863}$ = .3639
                                a
Multiply both sides by "a" and you get:

3,863 = .3639 (a)
Divide both sides by .3639 and you get:

$\underline{3,863}$ = a
.3639

a = 10,615 feet... or 2.01 miles... the distance from the observer to directly under the object.

Let us calculate the speed of the object. Let us assume that we know what direction the object entered the screen and departed the screen in a video. By using a compass, you can find the directions. Then, by going to Google Earth, you can go to the location and find the location of the observation and go 2.01 miles in the direction that the object entered the screen on the right. Next, go 2.01 miles in the direction that the object departed the screen on the left. Place a marker at both locations. Then, you can use the measuring tool to see what the distance was between the entry and departure points. Since you have a video, you can count the number of frames in the video from when the object entered, and then departed the screen of the video.

If for example 15 frames elapsed from entry to departure, and there are 30 frames per second in your video, that means that the distance was covered in ½ a second. Convert your distance to feet. Then divide the distance in feet by .5 seconds. That resulting number is the speed in feet per second. Let us assume that the distance that you measured in Google Earth from entry to departure on the screen was 2,000 feet. Divide 2000 by .5 and you get 4000 ft/second.

Google "converting feet per second to miles per hour". Plug in 4000 ft/second and you get: 2,727 mph!

Editor's Note: Since we finalized this book Missouri MUFON State Director Debbie Ziegelmeyer created a video analysis team which includes MUFON Field Investigators who are skilled at video analysis from several states. This team, of which Wayne Lawrence is a part of, is now finding multiple high-speed craft in video taken by witnesses who observed other UAPs at the time the video was taken, but were unaware of the Fast Movers until they or our team analyzed the video frame-by frame. In most cases this means looking at 30 frames per second, making the process time-consuming but productive. It is even more apparent after seeing many more videos that there are Fast-Movers in our skies and in our lakes and rivers at all times.

CHAPTER 2

THE FAST MOVERS: MY CONTRIBUTION

By Bill Spicer

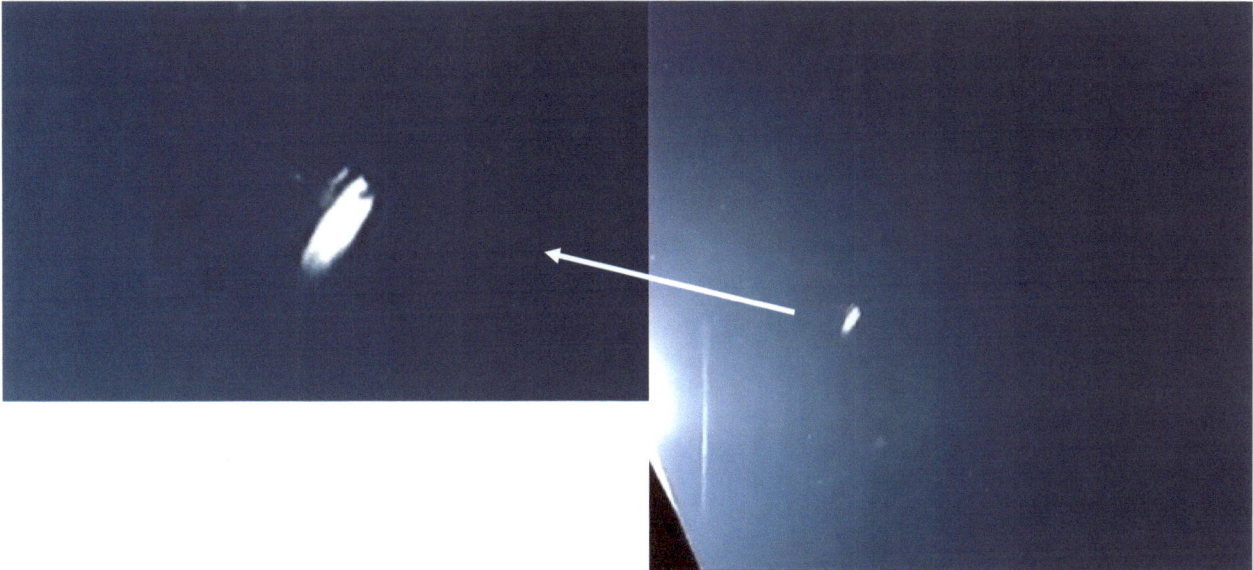

THEY ARE HERE – THE FAST MOVERS

Bill Spicer uses a fascinating technique of capturing UFO/UAP Craft with digital cameras and has an amazing story about his interaction with multi-dimensional Star Beings providing him the knowledge with which to see their craft.

He will be presenting a viewing method called *Quantum UFO Photographic Technique* used to take both video and still photographs of UFO/UAP Craft, showing that we have never been alone and are continuously surrounded by multi-dimensional Star beings here with their craft.

Bill has experienced a multiple of modalities during his life, such as a near death experience, missing time after witnessing a UFO/UAP Craft and has precognition of events in his life that have come true after his Star Being interaction.

His work has been reviewed by a known astrophysicist at the Harvard/Smithsonian Center for Astrophysics, who developed the gravitational lensing technique, and has made technical observations which support the Quantum Hologram theory tied to the Quantum Consciousness Field.

My Perspective

A new awareness and communication through dreams began for me after a lifetime of strange events began early in life and continue to this day. After a near death experience as a young adult I became aware that what we call existence is not always what it appears to be. We are constantly immersed in a Quantum Consciousness Field of energy that surrounds us. You only must accept this and a whole new reality will come forward to start a paradigm shift that will allow you to grow in your daily life.

Lucid Dreams

Have you ever experienced a dream that was so real every detail can be remembered, days and weeks or even years later? That happened to me in May of 2009. I became consciousness that I was not asleep, but very much awake and not in a normal dream, like most experience, but in a lucid dream. Soon it became noticeable that I was standing in the presence of a very tall being dressed in a white robe, emanating light, leading me around inside a Craft, while having telepathic conversations. During this experience I was informed telepathically that I would start seeing UFO/UAP Craft. Of course, I found this humorous and laughed at the thought, but got no response from the being I was walking beside.

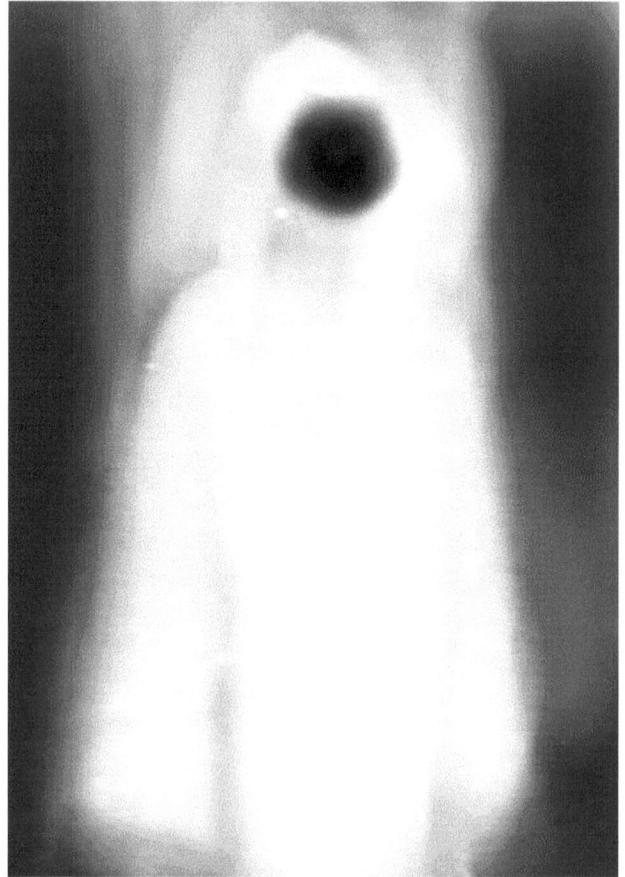

Drawing by Bill Spicer

The being was very much in a business attitude, giving me the impression that this was very important information and I needed to pay attention to fulfill the agreement I had made in the past. At that moment the telepathic images began to be downloaded into my mind, in an exchange with the being and then projected onto the wall of the Craft. This is when I began to take notice and understood that this was important to the being and began to think critically, trying to remember the details.

What I was shown at first did not make sense in a technical way, because I was shown a view looking back toward the Earth from a position high above it from space. From this position you could see the Sun back-lighting the Earth's edge and atmosphere, leaving only a small sliver of Sunlight.

Photo of multiple craft

Photo of white spheres and energy

From this position you could see a variety of white glowing spheres and energy moving about. I came away with the understanding that these are the UFO/UAP Craft, which just minutes ago I had been laughing about.

The Fast Movers were everywhere!

A New Reality

It was not even twenty-four hours later I was at work when the first sighting appeared in the Northern sky at mid-morning as a bright white light Orb that was not moving, providing me and co-workers who witnessed it an opportunity to discover if you were wearing polarized sunglasses you could see it. Without the sunglasses you could not. The bright white light was seen for over one hour before fading away. Each of us that saw the light were amazed and talked about it for the next week. Looking back

Image taken from video recording.

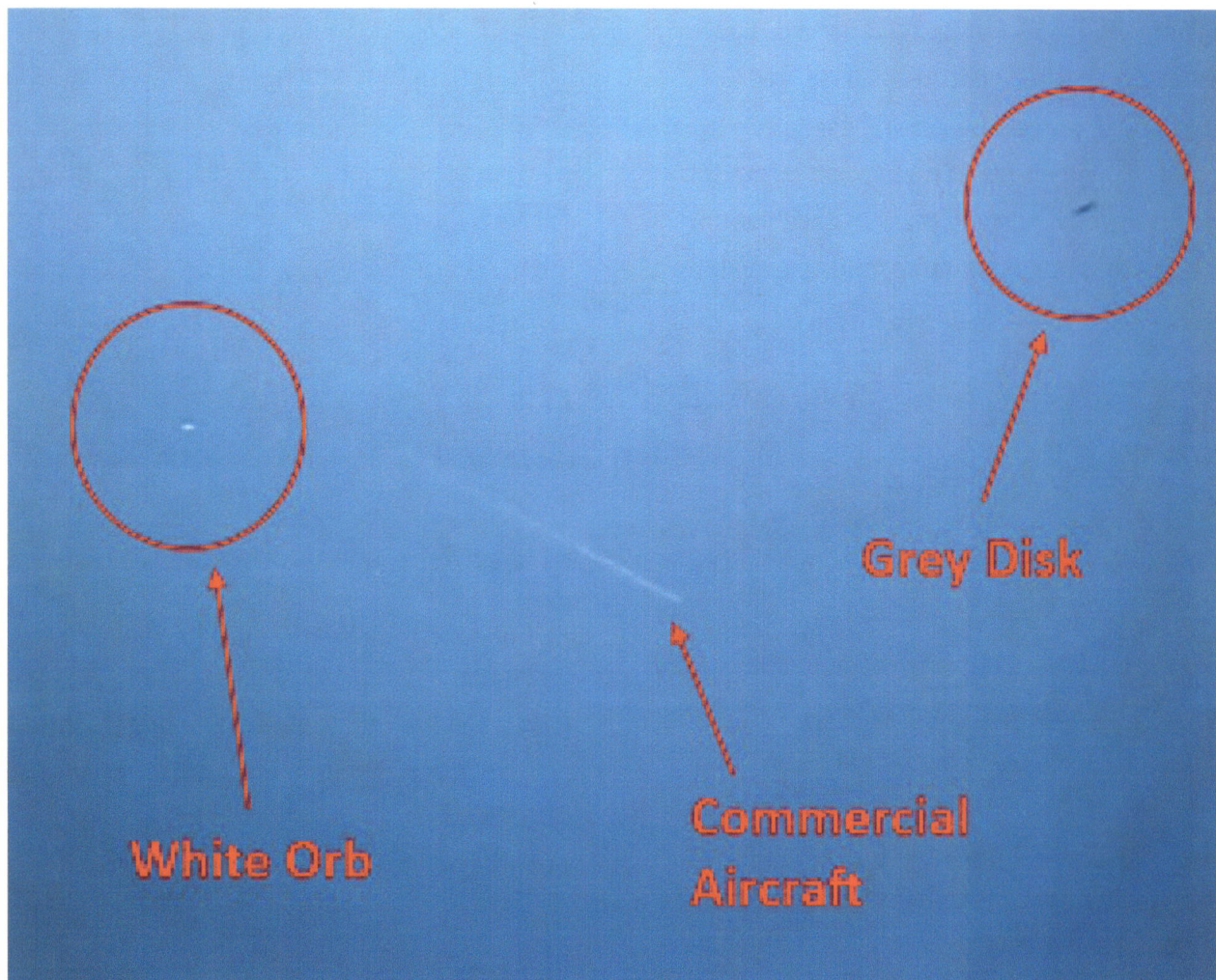

Image taken from video recording.
Notice the White and Grey Fast Movers. Photo: Bill Spicer

this was the first test of my ability to awaken and see through the first, second and third veil placed on humanity.

With the knowledge that while using a polarized pair of sunglasses you could see the bright white light, many daytime sightings began to happen. I purchased a small simple digital camera and began to watch the daytime sky for any bright lights such as Orbs either moving or stationary.

On May 30, 2009 my reality was finally shaken when my neighbor and I witnessed a bright white Orb moving about in the late afternoon

sky, but this time I had my small digital camera with me and made a video recording.

Imagine all the questions I had running through my mind. I would never have imagined this could be possible and happening to me. My neighbor and I are both technically trained and after this experience we began to change our sense of reality. We talked about this experience for weeks.

This began the start of my research into what could possibly be happening. Then things changed again.

THE LAST DREAM

Some months had passed and this time when I became very aware that I was back in a lucid dream with the same tall white robed being, there was much more telepathic communication.

This time I was shown again telephatic images about future events such as earthquakes on Earth and knowledge about the Craft I was on. As I started to ask questions, I was shown how the Craft operates, which is critical to understanding the ability to see the Fast Movers, while using a polarized lens fitted to a digital camera.

The Craft uses a propulsion system having the ability to be seen at a light frequency not visible to human eyes without using polarization filtering.

The process shown to me is called *Inflationary Vacuum State Propulsion*, which is used to propel the Craft, and will produce a light frequency or resonance wave of energy as it breaks the bonds of gravity, by giving off an energized electromagnetic field, creating a quantum-vortice of lattice ions forming a gravitometric field just outside the Craft. This forms a space/time curvature that is the same as gravity, while allowing the Craft to move at the speed of thought. The tall white robed being called it a thought projection. This corresponds to our understanding in quantum physics as *"Non-Locality"* and could be thought of as a *"Quantum Hologram"*.

A **"*Quantum Hologram*"** is defined as an extension of the known process of quantum emission/absorption. It is analogous to non-local quantum entanglement of particles, except it can pertain to matter of all scale sizes.

Entanglement – The state or condition in which an enduring relationship is created between atomic and sub-atomic particles during energy exchange or other processes.

Coherence/Quantum Correlation – The attributes resulting from entanglement such that wave-forms are aligned, and spins are correlated.

Non-Locality (Near and Far) - The transfer of such influences by thought or consciousness at the quantum level instantly, simultaneously and ubiquitously, through wave-like or field-like resonance irrespective of distance.

Resonance – The process of transferring and receiving influence and information non-locally.

The UFO/UAP Craft tend to move about their central axis, following the Right-Hand Rule for how electricity moves through a wire.

The Mechanics of Operation

The Craft have the ability to transfer energy from the vaccum of space and turn it into light having a frequency with properties at a given resonance are not seen in our 3D typically. This explains the glow and distortion of light seen around Craft during reported UFO/UAP Craft sightings, where space-time-gravity are being changed.

There are a series of Field Emitters localized around the parameter of the Craft that create the energized electromagnetic field. This can be seen as a distortion around the Craft as it moves or is stationary. The Craft have the ability to be thought projections based on what is known about quantum mechanics physics.

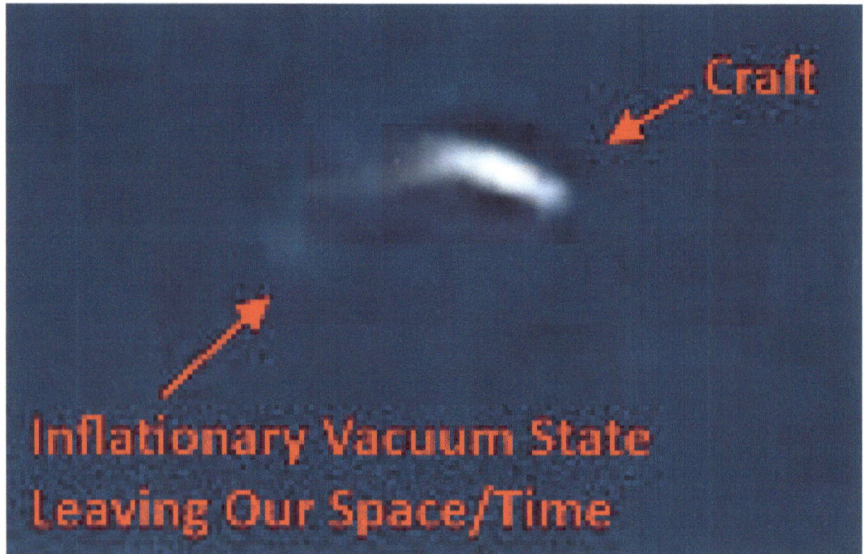

The ability to cross space/time is a real possiblity and has been shown to me over and over in photos and video taken of the Craft.

The ability to move from one location to another using only your thought as a medium is profound, but a real possibility with quantum mechanics physics.

When this happens, a hole can be opened in the local space/time as a craft can move into other local space/time or dimensions, which can be seen using the technique called **Quantum UFO Observation Technique (QUOT).**

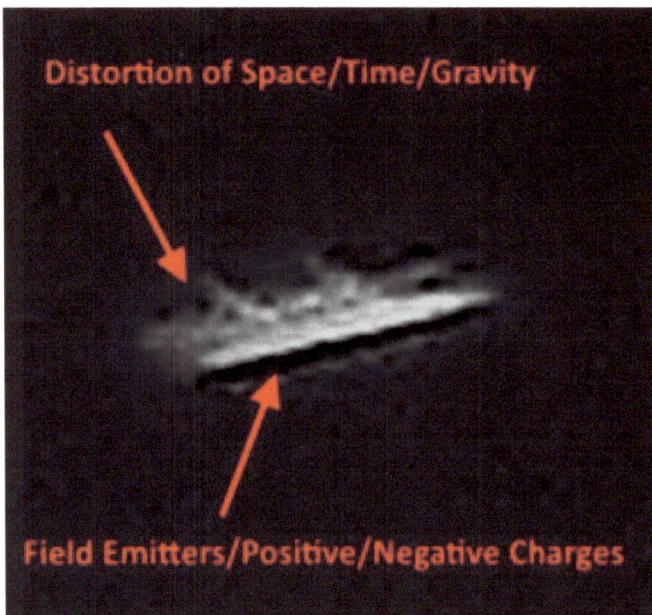

Left: Craft (black & white for clarity) displaying the operation of Field Emitters, distorting space-time-gravity around the perimeter of the Craft

THE SETUP

As you begin the process of capturing UAPs in our skies, any digital camera can be used, but the better the camera, better the results. I also recommend using a tripod when available to steady the setup. It is highly recommended.

The cameras I use are not very complicated and I use the AUTO-MODE settings at all times.

Using this process, I have instructed people who had no better camera than a cellphone and they have had remarkable results taking photos and videos to see the craft. For some it was their life changing experience known as *Awakening.*

Steps involved as I was shown.

Step 1 –

(See right) Position a HD digital camera on a tripod in the shade of a building or structure.

You will need to acquire a pair of Polarized sunglasses or a polarized lens made for your camera before continuing. Then you can complete the setup. A dark shaded polarized lens will give you an excellent result.

Canon DC420

Kodak PIXPRO Model AZ501

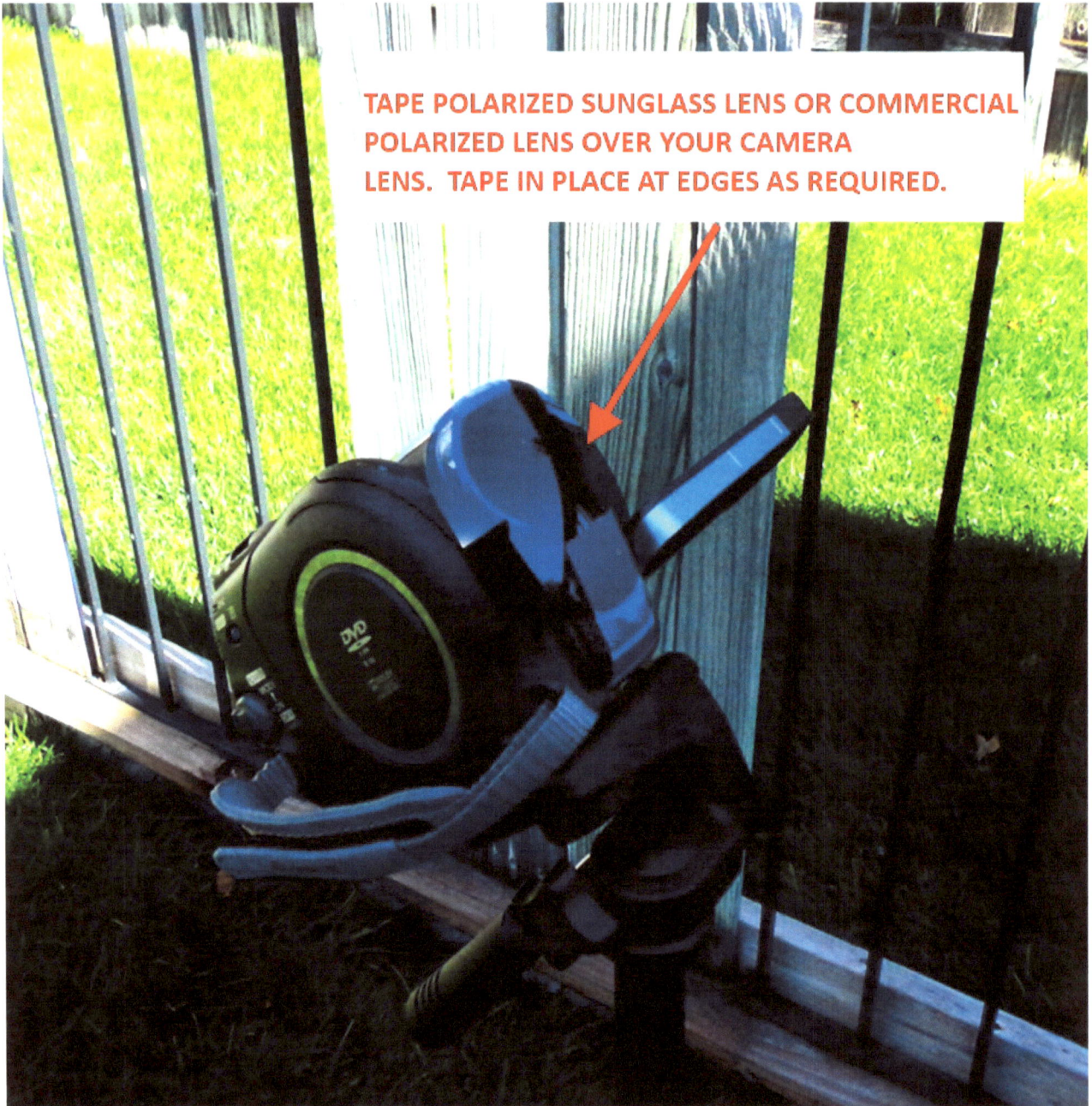

TAPE POLARIZED SUNGLASS LENS OR COMMERCIAL POLARIZED LENS OVER YOUR CAMERA LENS. TAPE IN PLACE AT EDGES AS REQUIRED.

Step 2 –

Attach or hold Polarized Sunglasses over the camera lens, or attach a commercial polarized lens. I recommend removing a single lens from a pair of Polarized Sunglasses and place it over the camera lens. Tape edges as required.

Polarized lenses have a special chemical applied to filter light. The molecules are lined up specifically to block some of the light from passing through the lens. The filter creates vertical openings for light so that all of the horizontal light waves are blocked. Objects are darker but clearer and details are easier to see. Not all sunglasses are polarized, so check the label before purchasing.

Step 3 –

Aim camera towards a building or structure and focus on the edge of the structure. Move tripod towards or away from building or structure as much as required.

CAUTION

NEVER LOOK AT THE SUN DIRECTLY WITH YOUR EYES. USE THE CAMERA VIEWER ONLY TO LOOK THROUGH LENS WHEN AIMING.

AIM CAMERA IN THIS AREA ABOVE STRUCTURE.

Step 4 –

Complete focusing camera to FULL ZOOM once the sun looks like a small thin edge of light. The UFO/UAP Craft will appear above this edge of light from the sun.

I suggest using HD video movie mode to be able to record and see UFO/UAP Craft, if your camera has that feature. They appear and disappear quickly and may require a video editor to see or still frame the Craft, to create digital prints of your favorites.

Have had success using the continuous still photo application found on most digital cameras and in addition all my cameras the AUTO MODE setting will get the best results.

I have found it only necessary to take 10 to 15 seconds worth of video at a time to begin seeing the Craft appear. Most show up in the first 30 second to 60 seconds after starting a recording session. There have been times where I have taken five minutes worth of video and found many UFO/UAP Craft in the field of view.

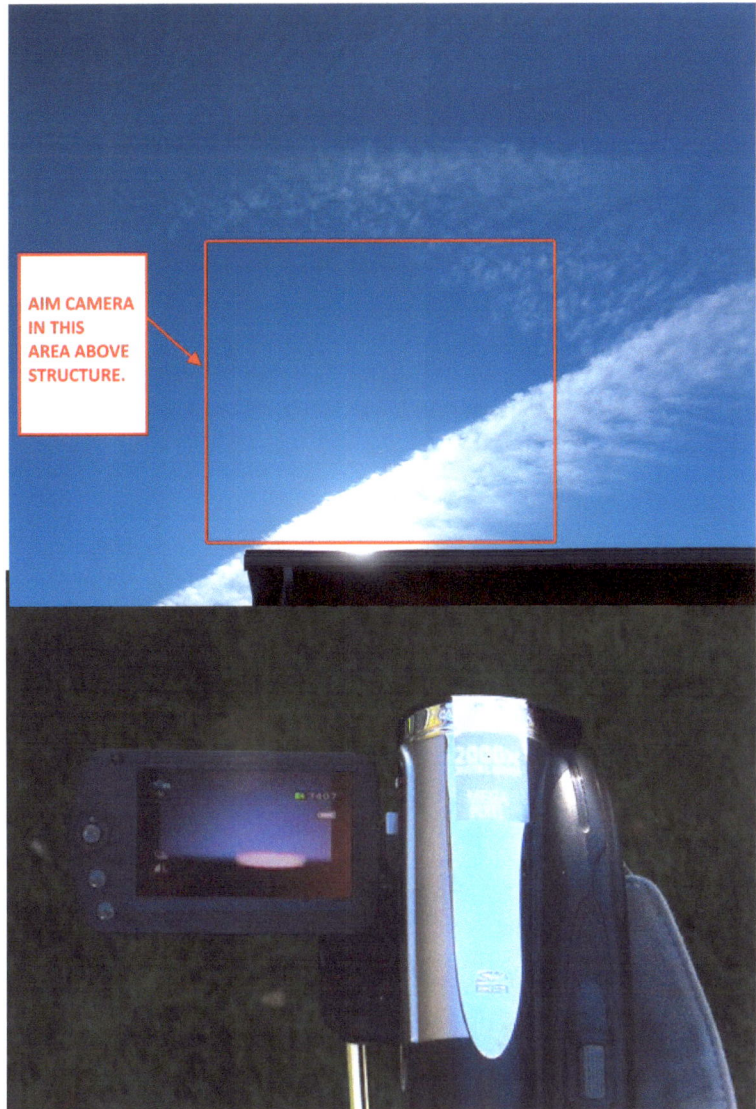

Camera Settings

- Picture Quality—16 Megapixel
- Special Affects—OFF
- AUTO White Balance—ON
- 50x Optical Zoom

Mental Preparation - Telepathic Messages

This is the step that I had the most to learn about. The ability to telepathically communicate asking that the beings appear and show themselves was not at first what I expected. It is like being respectable. Yes, the beings can understand your language or thoughts, especially your visual thoughts.

Here are the steps I use and have taught others to use with great success.

Step 1- Go into meditation before going outside to begin setting up the camera. Mentally invite the Star Beings to show themselves. Your visual thought projections should show where you are on the planet in a visual way. Just be friendly. I have found there is nothing to be fearful of when trying this technique.

Step 2 – Remember contact can come in many forms, but most will show you their craft. Other forms of contact may be by Lucid Dreams, a telepathic message or strange electrical disturbances.

Step 3 – Wait 5 – 10 minutes after setting up the camera and the craft will begin to show up for a first time experience. Please be patient if this is the first time to prepare. They may not show up on the first attempt, but keep trying.

This craft appeared after I meditated and requested the beings to show themselves

Camera Settings for Video

- Professional Mode

- Picture Quality 1152 x 864 Special Affects—OFF

- Shutter Speed 1/120th Second

- Light Sensitivity - Adjusted to highest white balance setting as measured form Cumulus Cloud tops

- 38x Optical Zoom

THEY ARE HERE

From this point on you should begin to expect to see many different UFO/UAP Craft. Here are many examples of Craft I have seen since May 2009 using the *QUFO Technique*.

This selection of photographs are screenshots taken from video using the Kodak PIXPRO camera:

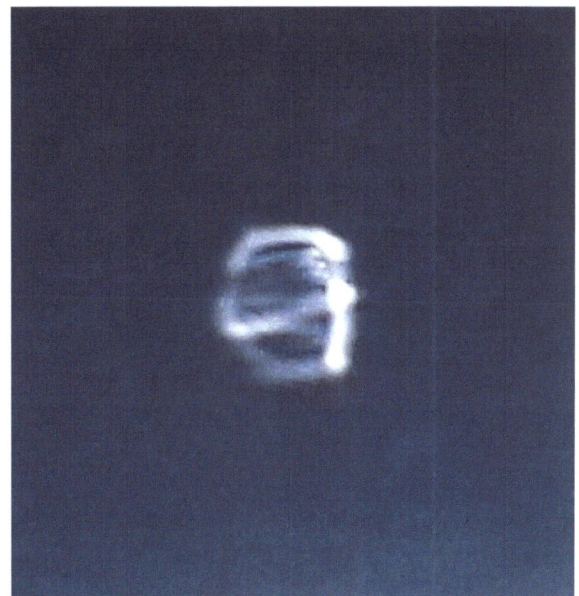

More photos taken with my Kodak PIXPRO camera:

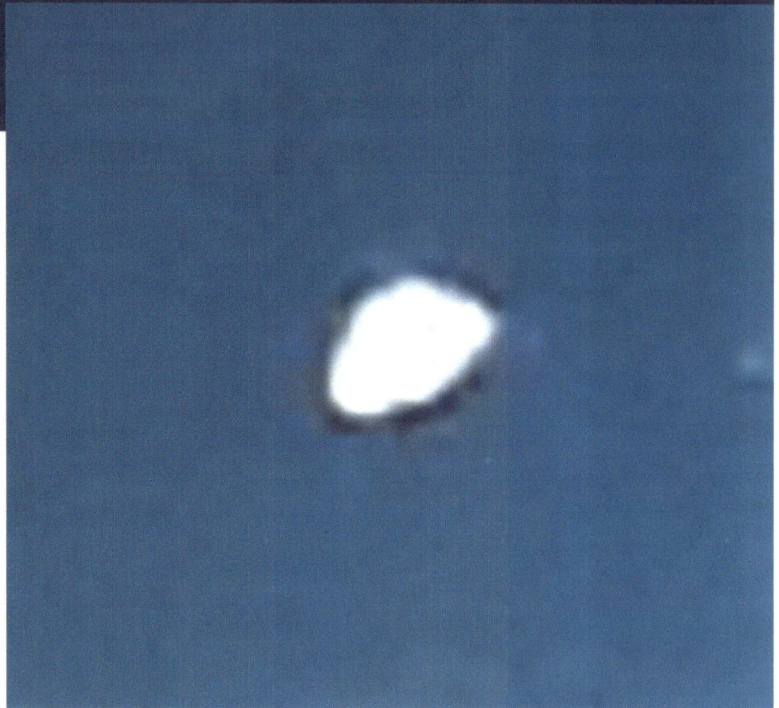

Close-up of object in photo above

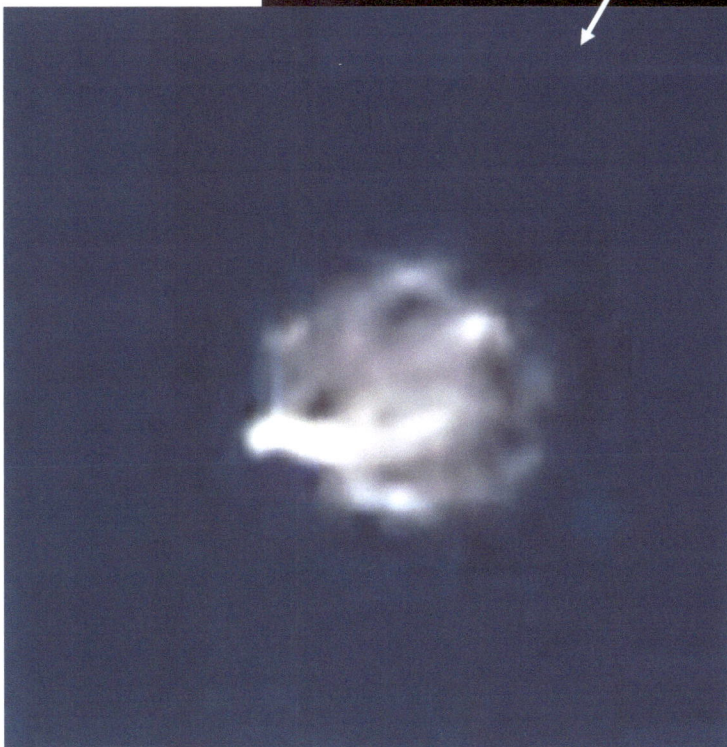

Close-up of object in photo above

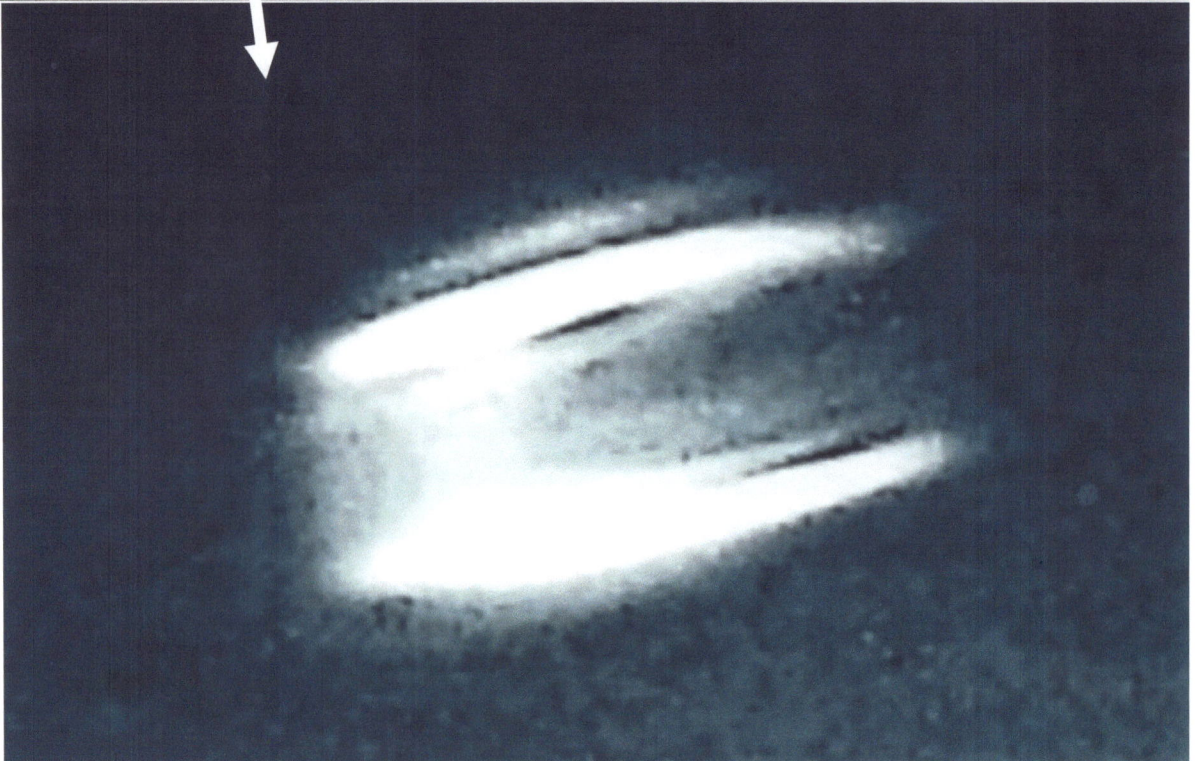

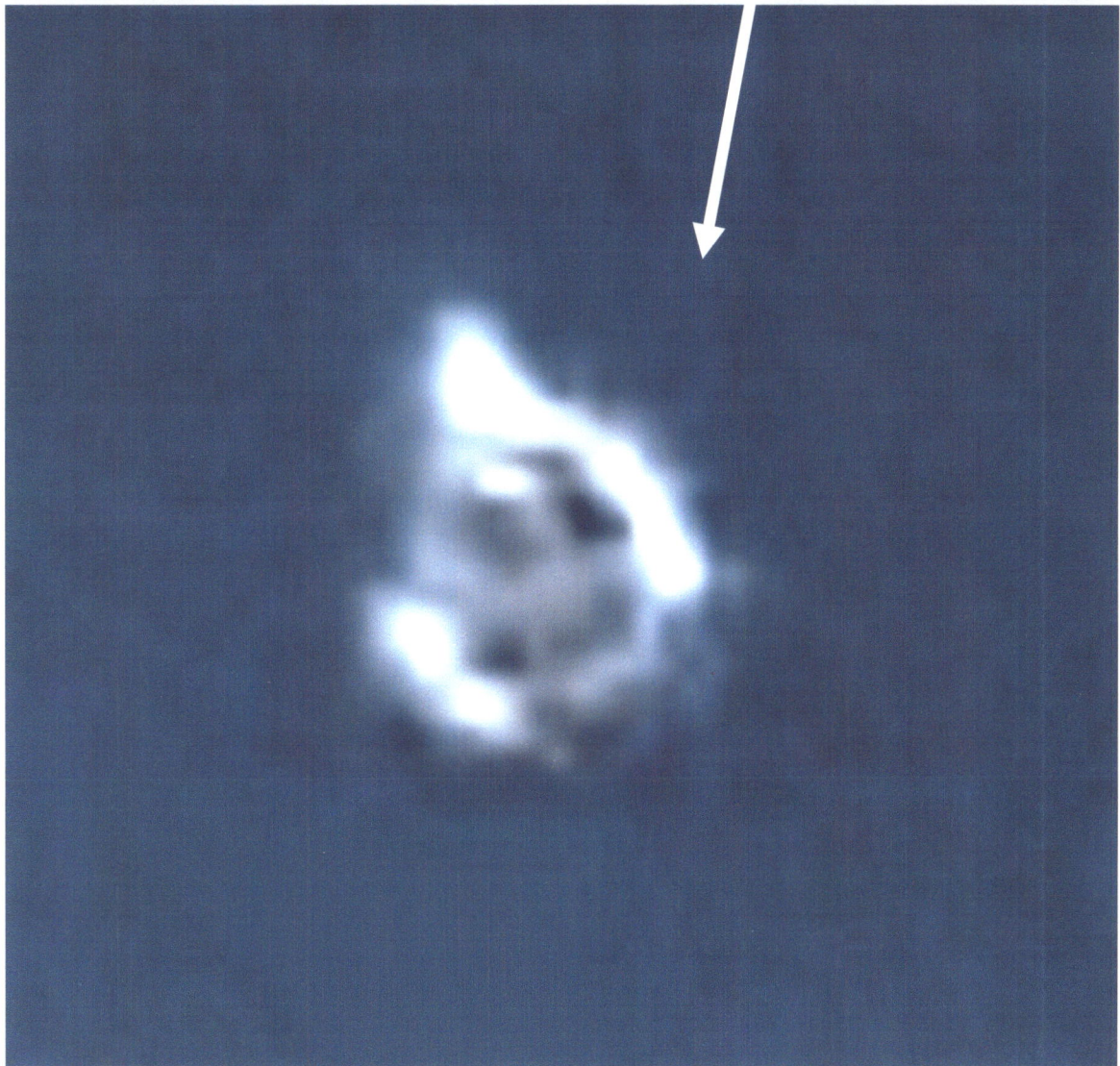

Right: Close-up of object in photo above.

Editor's note: This exact same object has been captured by other witnesses in different locations.

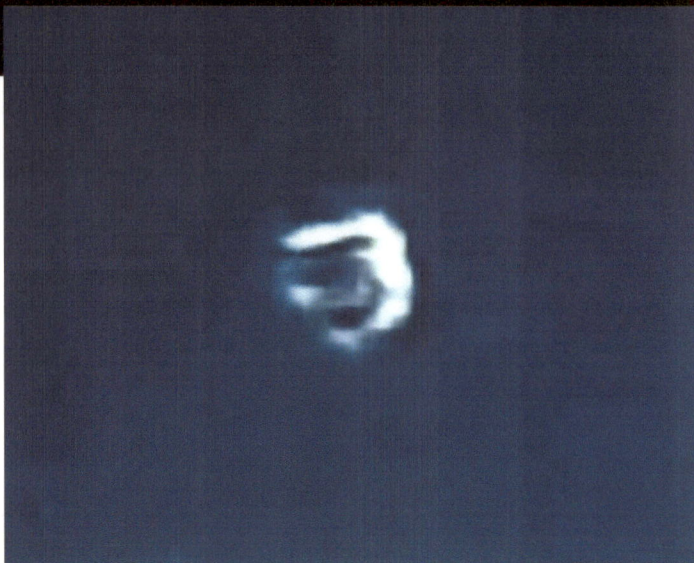

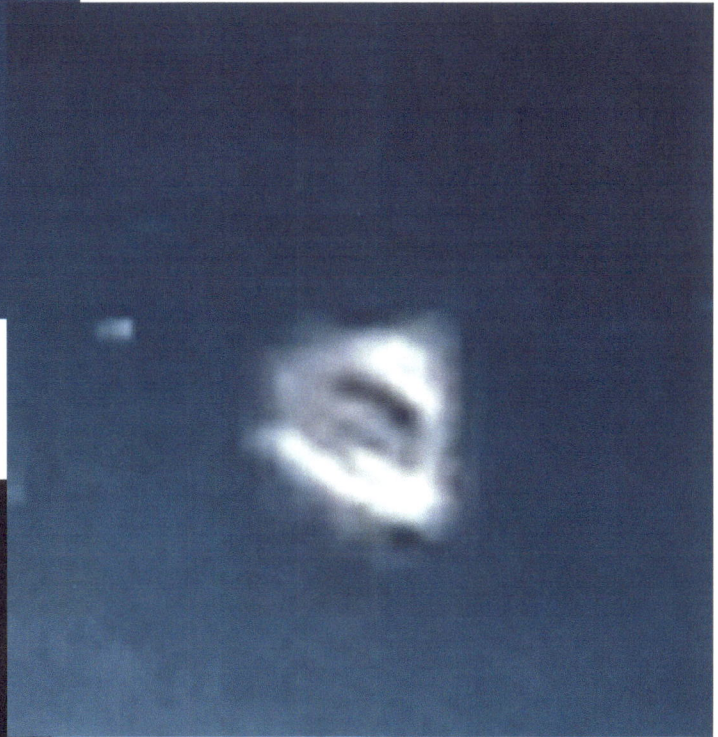

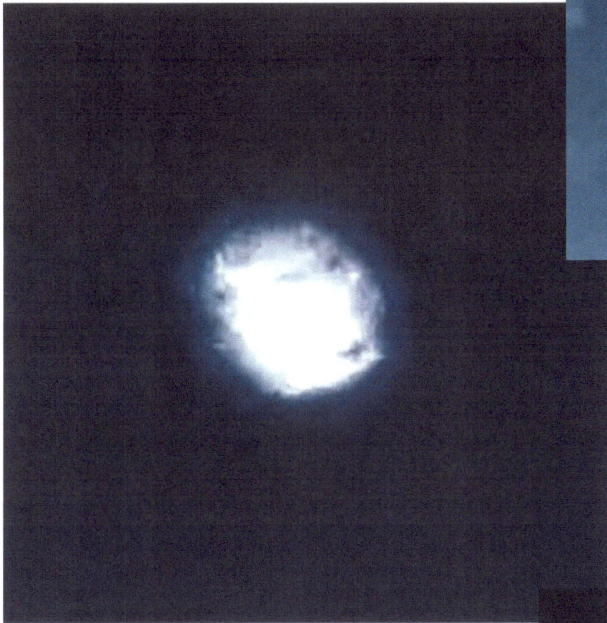

Close-up photos of more objects captured with my camera.

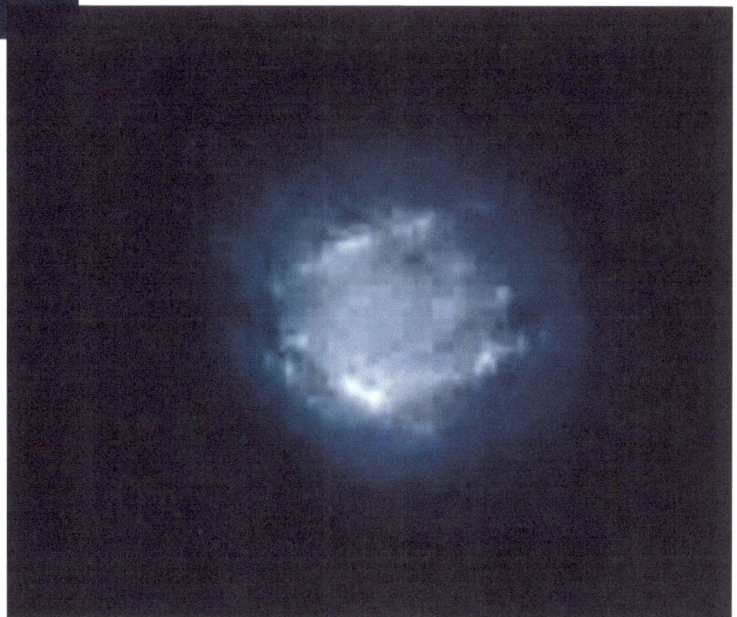

This selection of photographs were taken from stills using the Cannon camera:

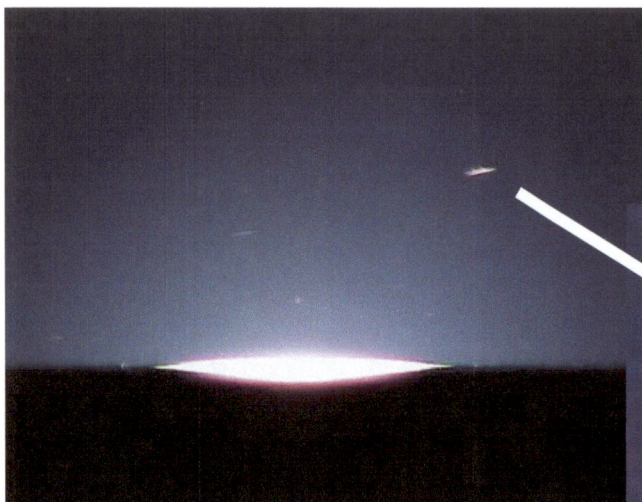

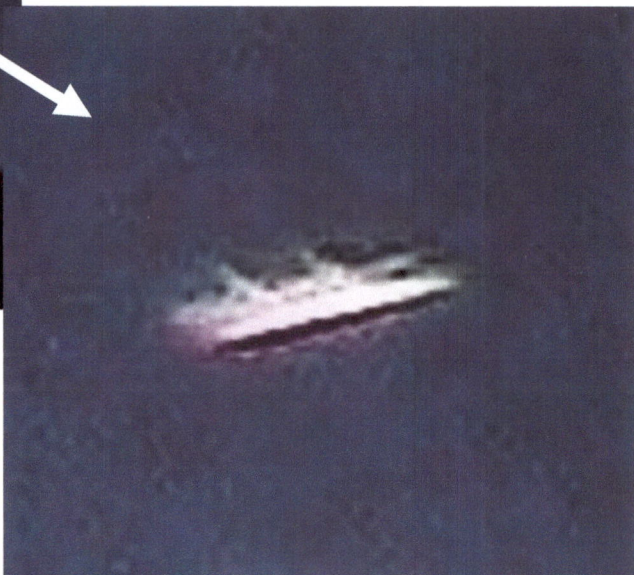

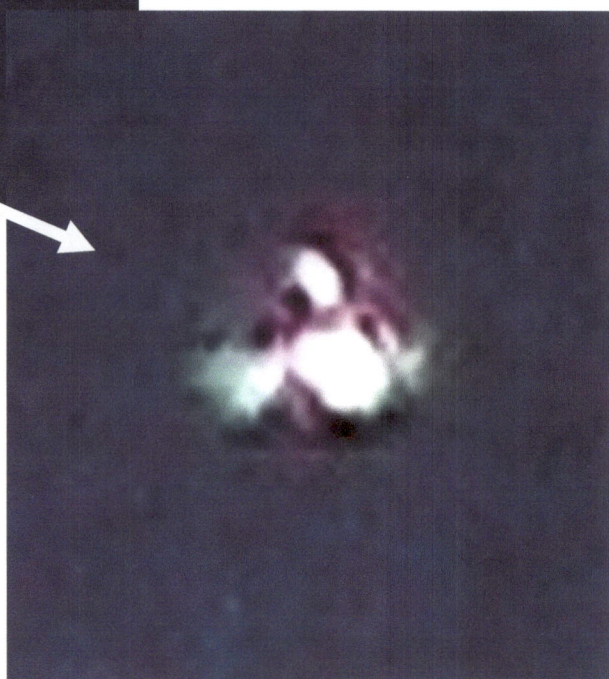

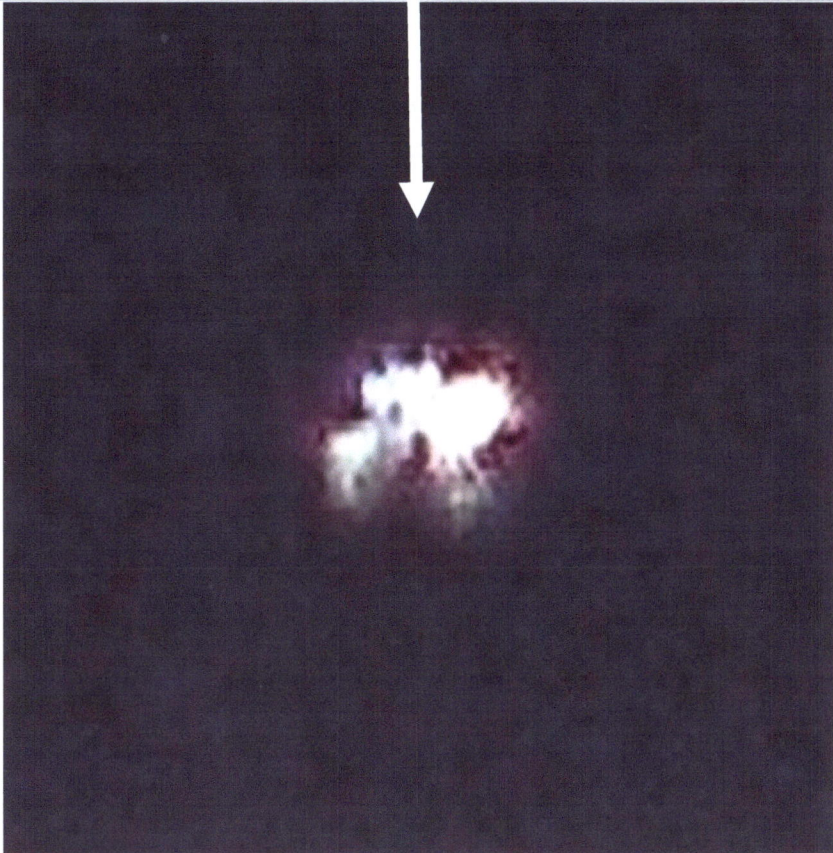

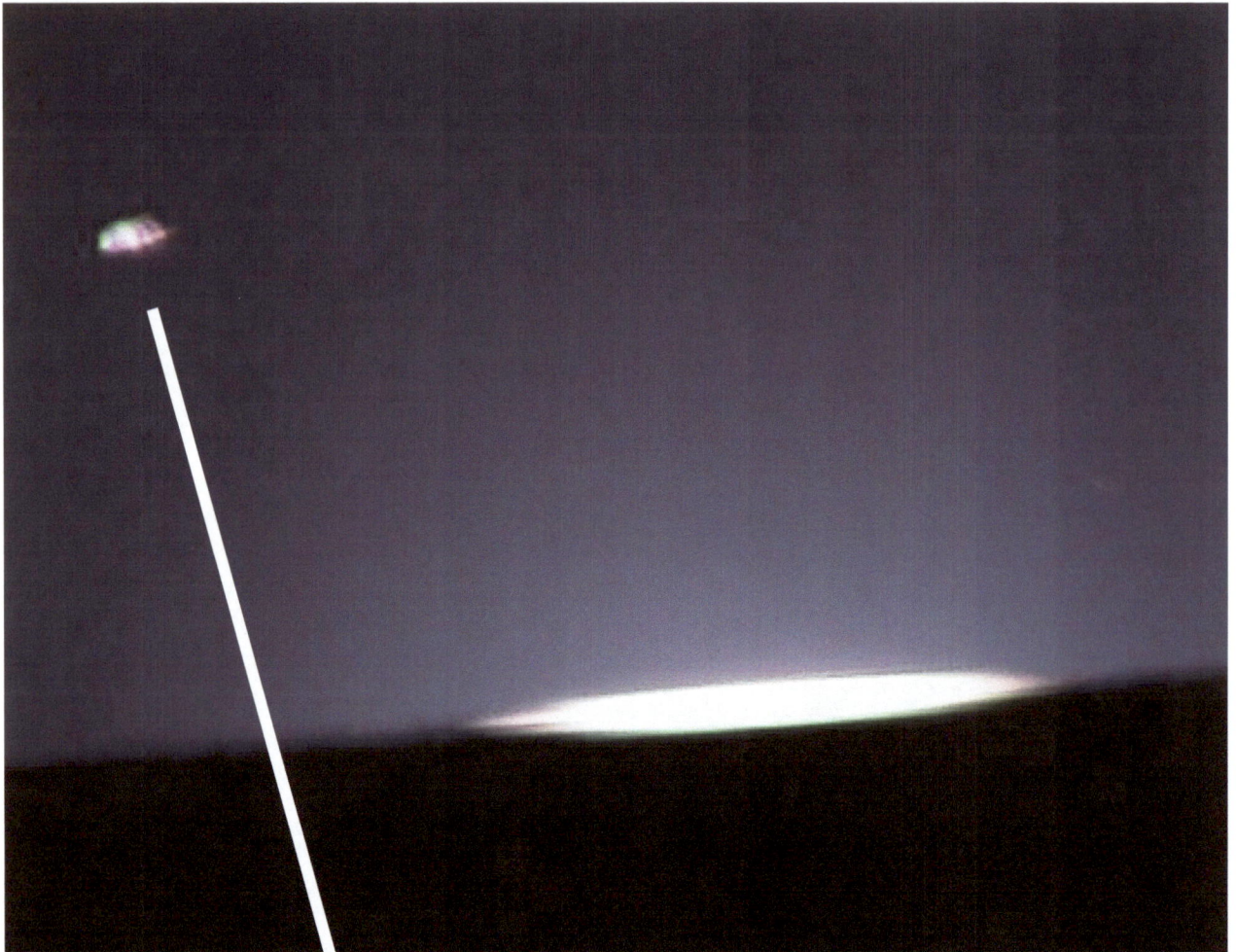

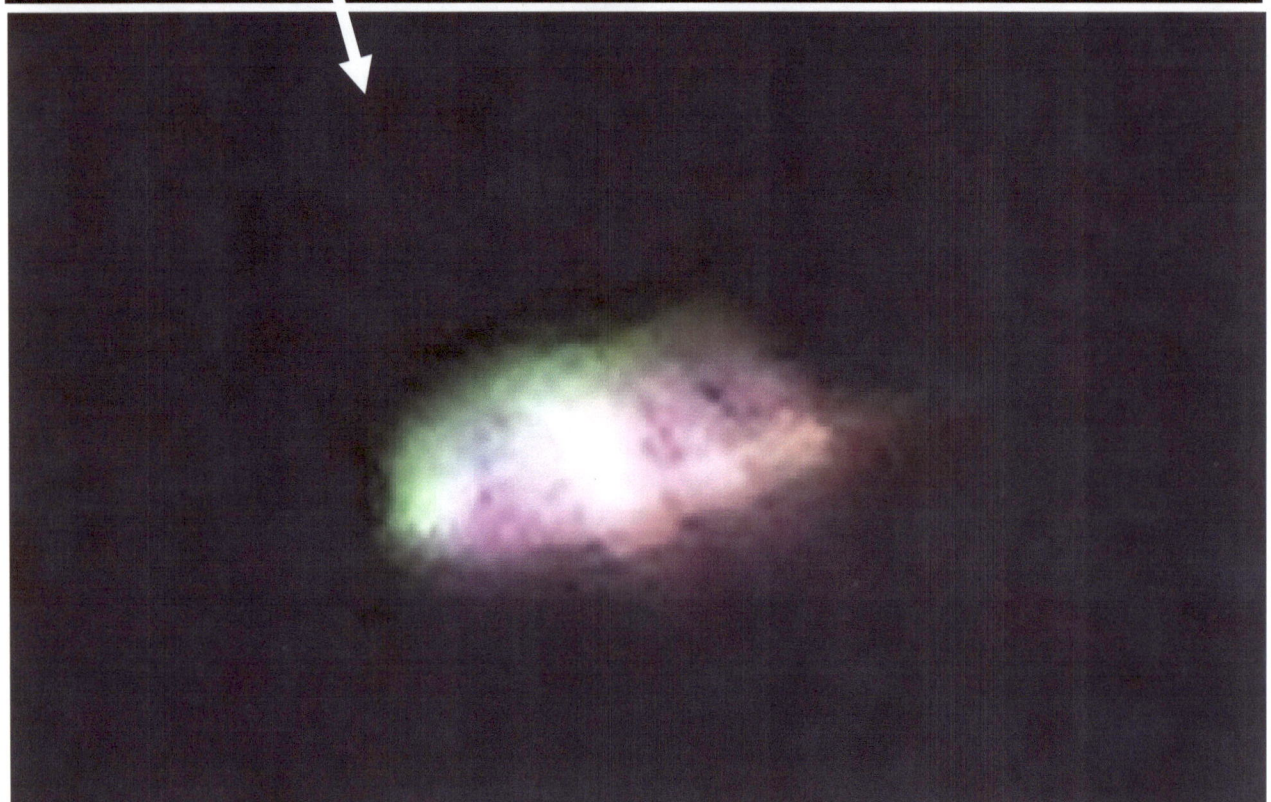

Other Sightings

As you progress using the mental preparation steps, the Craft will become visible even when looking away from the sun. This is when they know you are fully awakened and have mastered the QUFO Technique.

Using a cellphone digital camera and polarized filter expect to see examples such as these, where the craft come over your position during daylight. I call these the Fast Mover – Watchers.

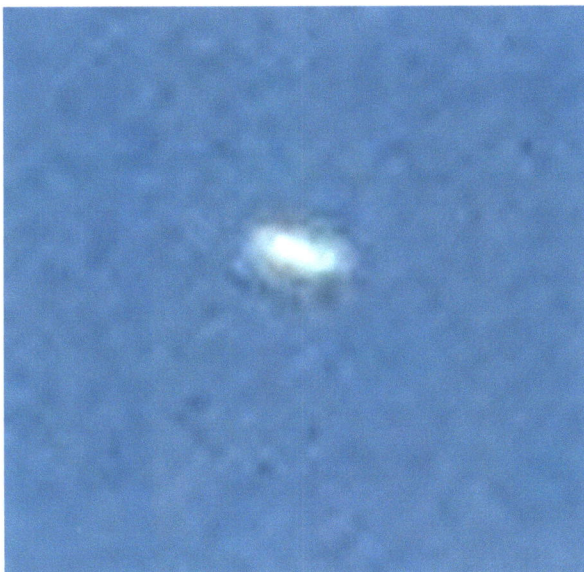

Craft seen over my home not moving, once I asked using mental telepathy for it to stop.

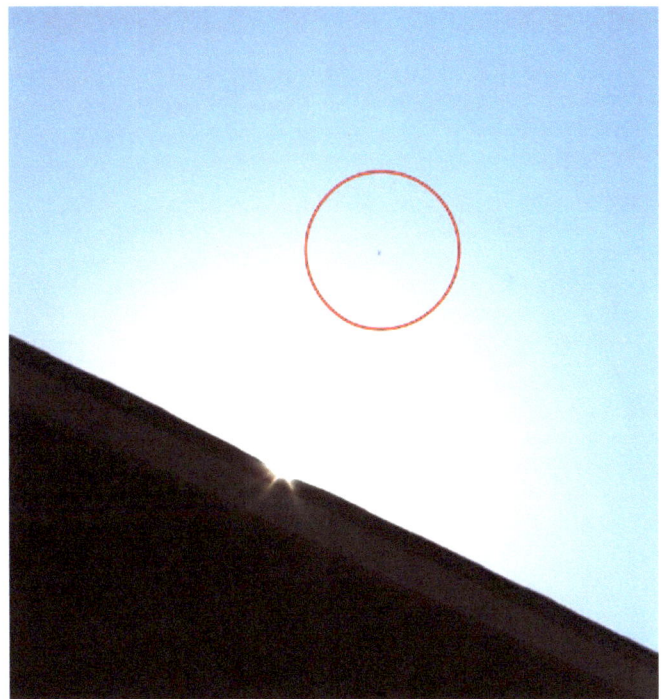

More craft were observed stationary over my home

This craft came over my home in daytime. The large craft was estimated to be thirty feet in diameter

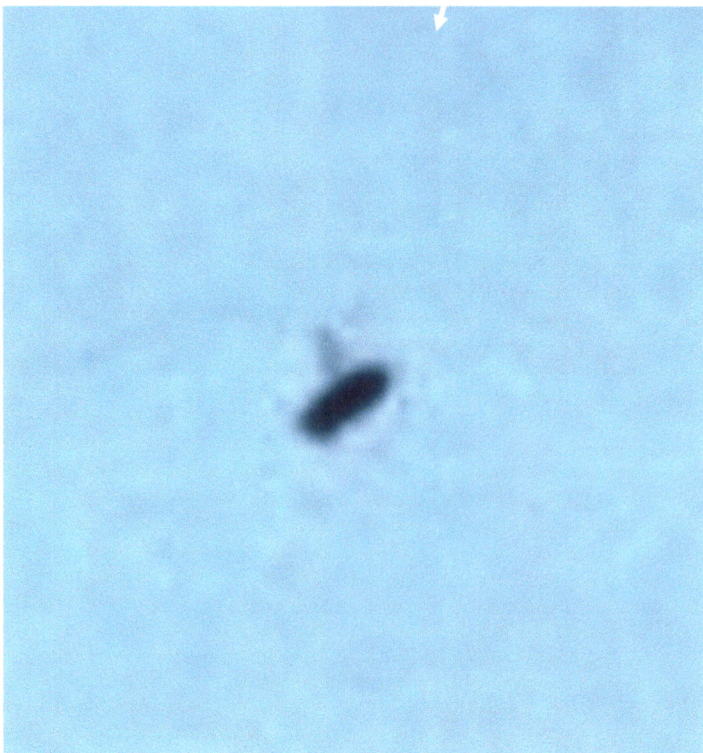

Close up enlargement of large craft as it moved, displaying a quantum-vortex on top.

PHOTO EDITING

Most photos and video will need to be edited to enhance or turn down the luminosity due to the bright background experienced with allowing the sunlight to be present, even with the filtering. I use a variety of different commercially available software that is available today.

Sometimes it is easier to see the contrast in shapes if photo editing software, such as paint.net, is used to turn the image to black and white. This increases the contrast where by details can be analyzed.

There are free and commercial video viewing and editing software available. From my experience you will want one that will allow for screen shots from where you can capture and analyze the video and movements made by the Craft – Fast Movers.

Cropping and enhancing the photos are easy and enable you to show in greater detail what the Craft look like. Many times, you will see that they show up very clear and without the distortion caused by the energized electromagnetic field, also causing the outside surface of the Craft to appear as glowing.

Right: High contrast with color will show great details.

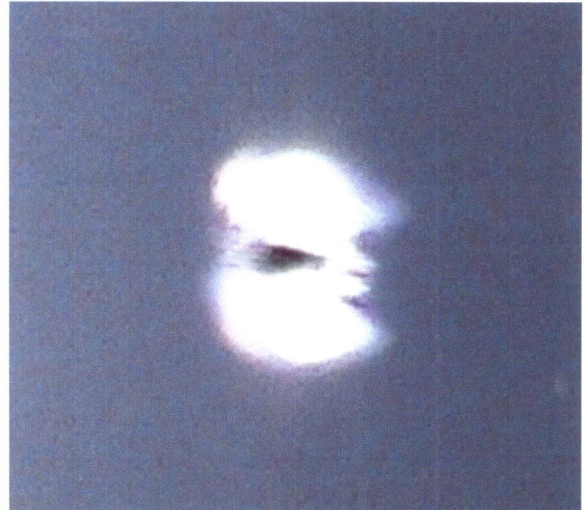

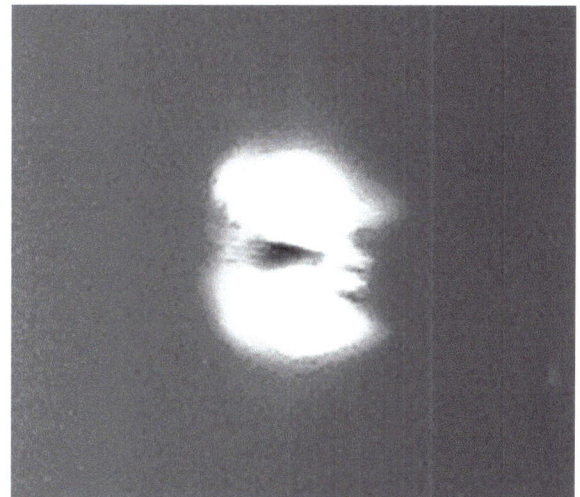

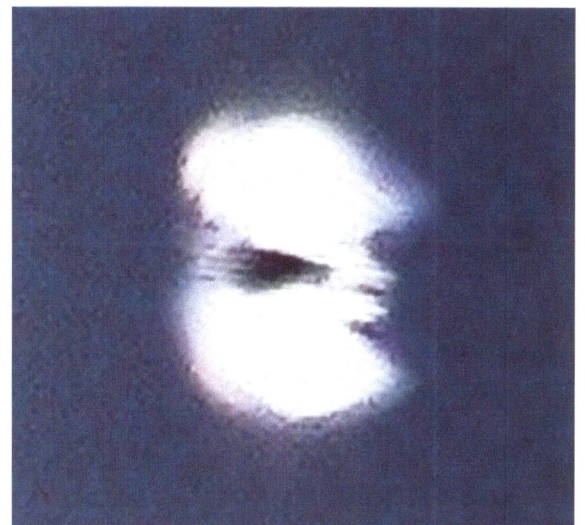

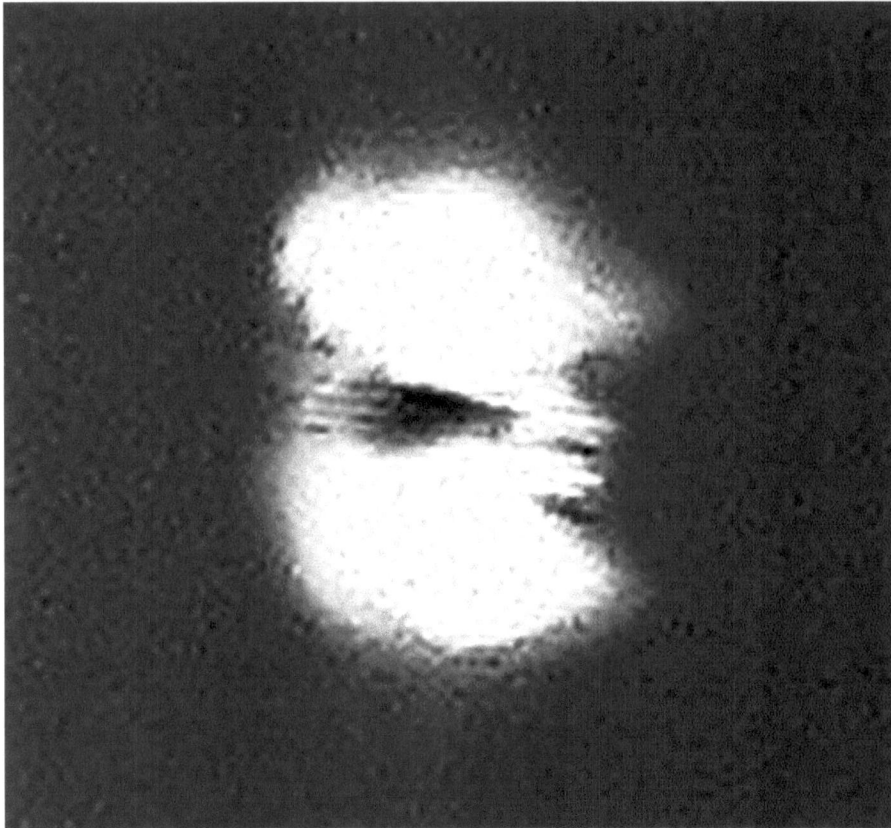

Higher contrast and Black and White stills will show details that are amazing

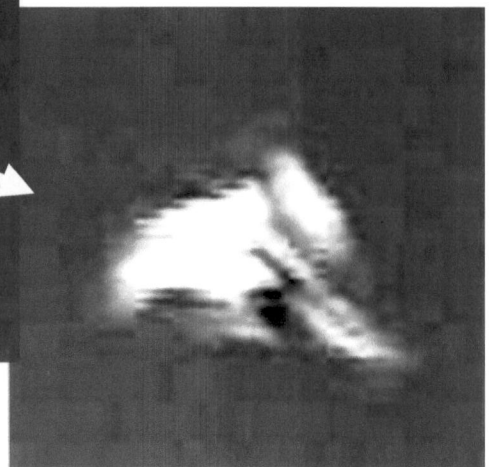

CONSCIOUSNESS

At this point you might be asking how it is possible for the craft and beings to travel through space/time. It is possible using quantum mechanics.

When this "observer effect" was first noticed by the early pioneers of quantum theory, they were perplexed. It seemed to do away with the basic assumption behind all science: that there is an objective put out there, irrespective of the human. If the way the world behaves depends on how – or if – we look at it, what can "reality" really mean?

Example: Is light a wave or a particle? It is both. It depends upon whether a conscious mind is present. The now famous [1]Double-Slit Experiment comes to mind.

There is a quantum consciousness field that exists as the beings showed me and it has properties that sound unbelievable. It is a form of light at a specific frequency and when you touch it while in your astral body, allows you to travel through space/time. The soul or astral body are very much inter-twined with the physical world, but are not held as part of it.

With this knowledge it allows the craft with its beings the ability to travel in a very efficient

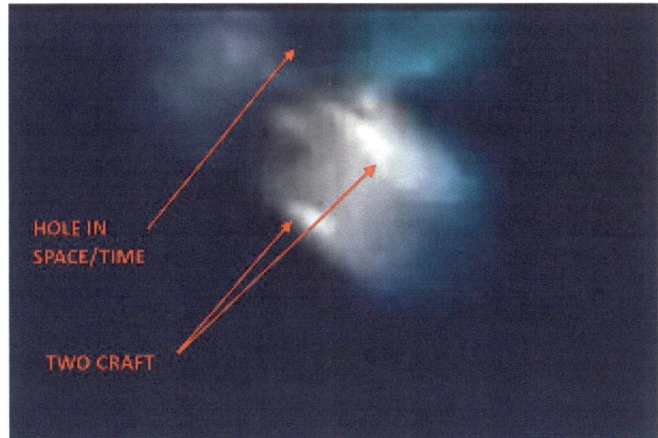

way to the far reaches of existence - or multi-dimensional existence.

Pictured above are two craft coming through our space/time leaving a hole in it, while using visual thought while linked to the quantum consciousness field, from where all creation derives. It is like jumping through a portal or wormhole. The beings visually think and guide the UFO/UAP craft to where they want to go.

This field, "sentience or awareness of internal or external existence", is at play within the Quantum Vacuum as a part of the properties of nature, that include harmonic frequencies of light, or thought projections (holographic) coming through the veil. By doing so, it allows the observer to interact with other conscious, non-human intelligent beings and their craft, that exist outside our three dimensional space -time frequency.

*(1) In modern physics, the **double-slit experiment** is a demonstration that light and matter can display characteristics of both classically defined waves and particles; moreover, it displays the fundamentally probabilistic nature of quantum mechanical phenomena. See https://en.wikipedia.org/wiki/Double-slit_experiment for more information.*

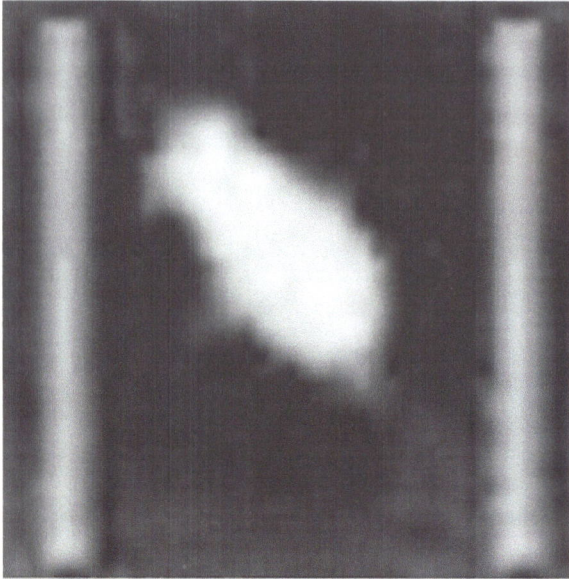

U.S. Navy F-18 Hornet Gun Sight Image
Taken 2004 (Off San Diego, CA)

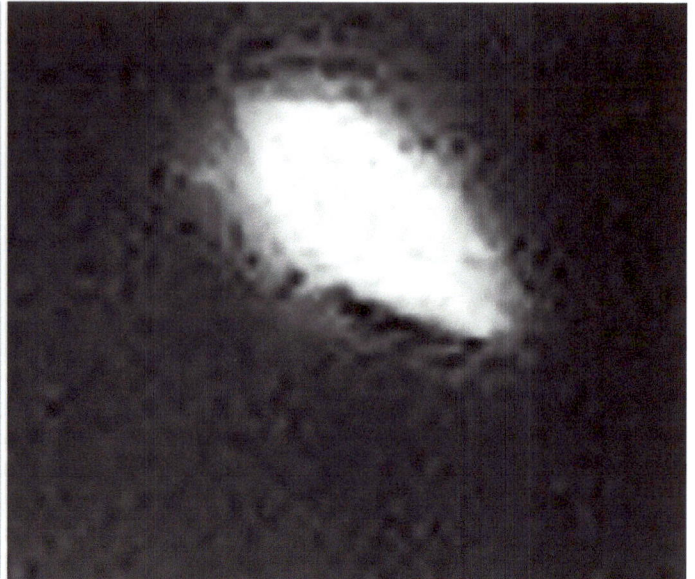

Bill's Photo - Taken 2012

BOTH LOOK SIMILAR IN SIZE AND SHAPE

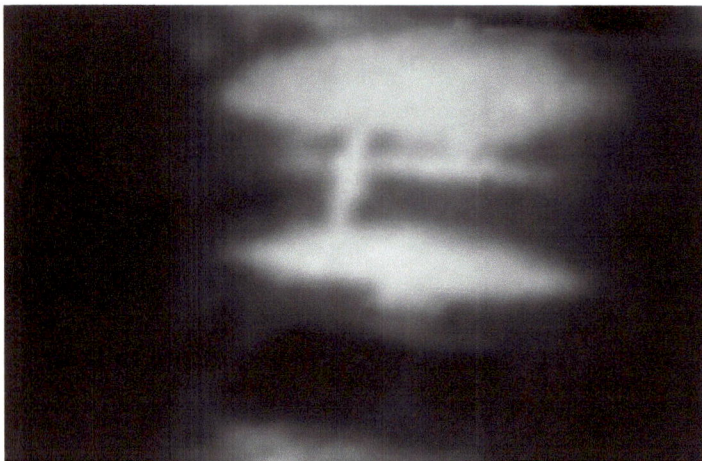

Boyd Bushman's UFO's
at Takoff.

Bill's UFO in Flight.

BOTH LOOK SIMILAR IN SIZE AND SHAPE

Conclusion

There is a bold new reality to explore and you do not have to travel very far to engage it. It could begin right in your own backyard where you can create your own X-Files collection.

Yes, disclosure is happening. Even experienced professionals in aviation and governments are finally coming around and saying that we are not alone.

For example, on December 17, 2017 the world found out that there were videos released showing navel aircraft engaging a white disk or Tic-Tac shaped UFO/UAP Craft. This was no joking matter according to the aviators who experienced seeing the Craft. The Craft were detected by military radar and Infrared cameras and each one very much looks like the type of Craft I have seen since May 2009.

There are other examples of Craft pointed out by the late government scientist Boyd Bushman and during his last interview made references to Area 51, where he discussed how the UFO/UAP Craft are constructed and exotic metals used. He also pointed out some of the origins of where the beings come from.

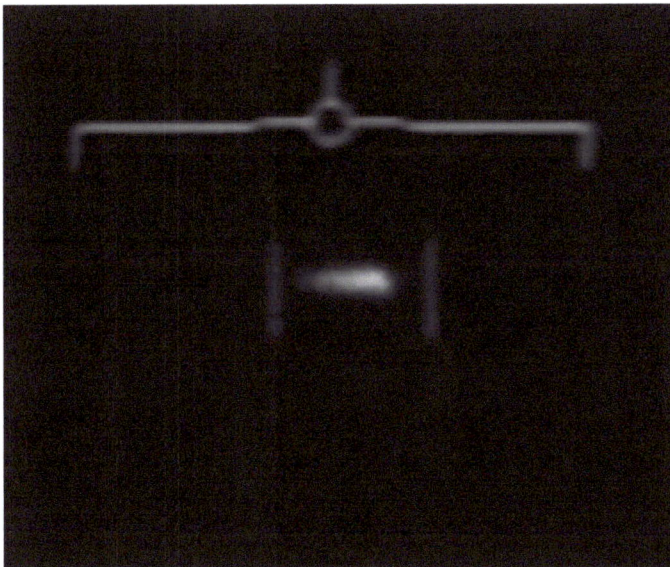

FLIR1 Official UAP Footage U.S. Navy - 2004

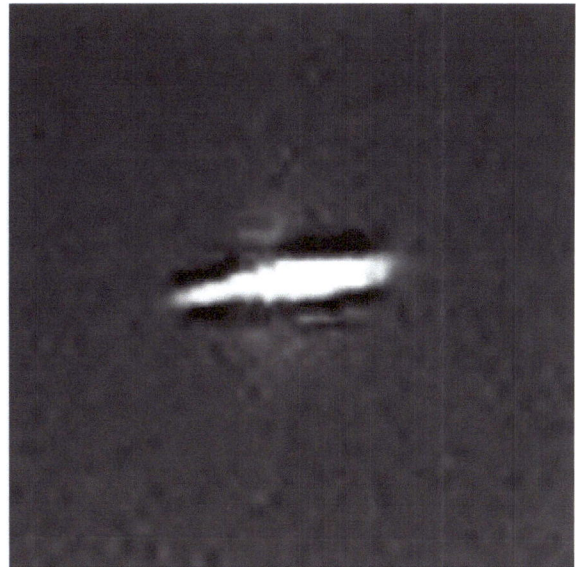

Bill's UFO - Sept. 23, 2018

Similar Outside Profiles

A Quantum Hologram?

On March 29, 2020, I went outside with my camera and asked the question: "Do we live in a quantum hologram"? A MATRIX? The answer implied YES.

Below: I aimed my camera into the sky, but without zoom and using my polarized filter. A whole chain of Light Ships (My Guides) appeared out of nowhere, all in alignment and facing the camera (me). They also show up in the foreground. These are not fractals or lens flares. They appeared after I had the camera aimed.

2020/03/29
02:53

Original photograph. The light ships are to the left of the sun in a line. A close-up of the ships is on the next page.

By Margie Kay, Wayne Lawrence, and Bill Spicer

Close-up of the light ships which appeared after a question I posed on March 29, 2020

About Lens Flares

Lens flares occur when a camera is aimed at something bright such as the sun or a streetlight. Flares manifest in two ways: as visible artifacts, and as a haze across the image. The haze makes an image look washed out. The double-lens of the camera traps light between the lenses and causes the appearance of a circular or foggy object in the photo. The circular shapes are due to the shape of the iris. Other shapes cannot be obtained, therefore, the objects in these photos are not lens flares.

Photo of Lens flares

Credit: Hustvedt / CC BY-SA (https://creativecommons.org/licenses/by-sa/3.0)

84

DISCLOSURE IS HERE

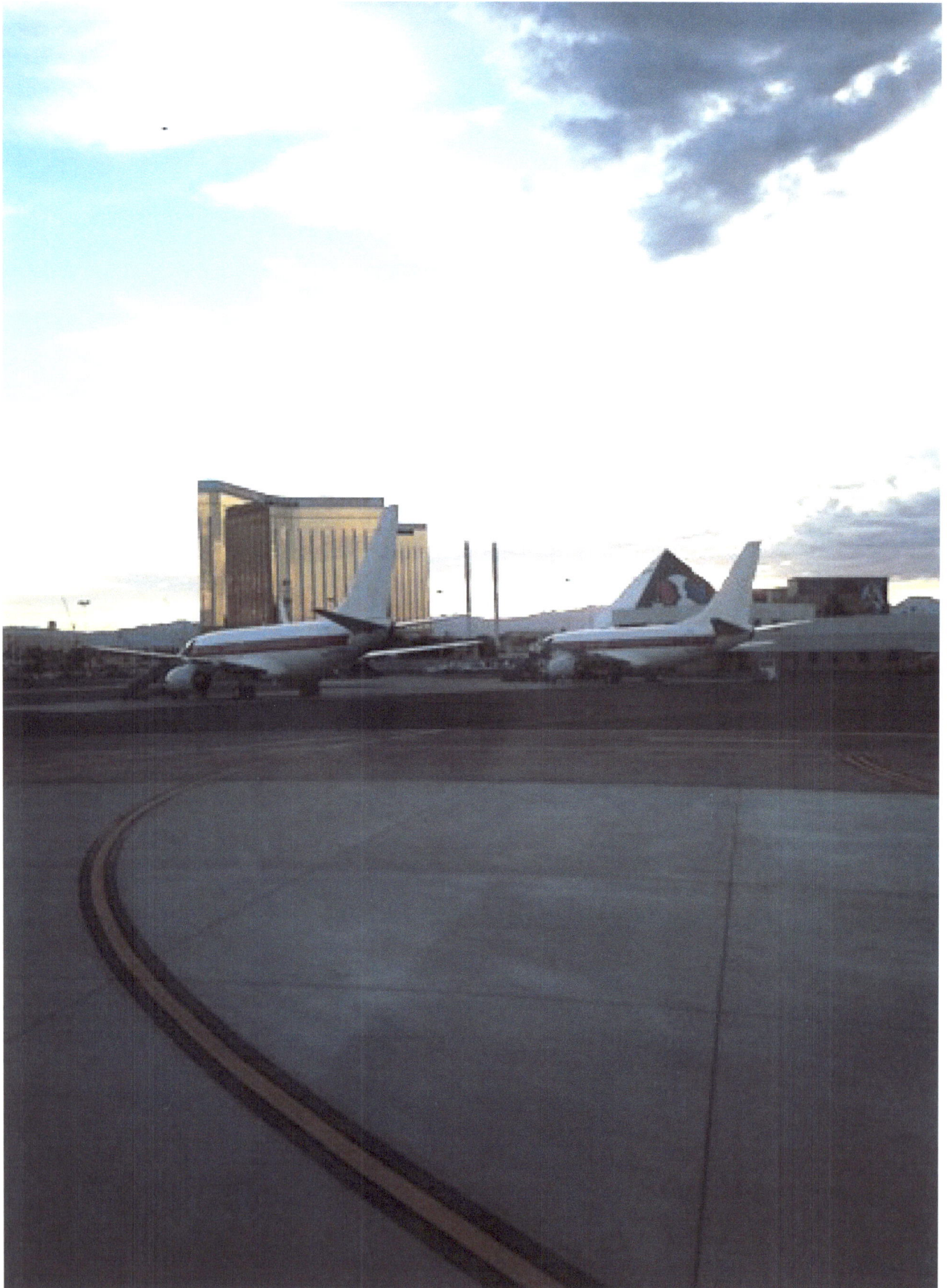

THEY ARE HERE

THE FAST MOVERS

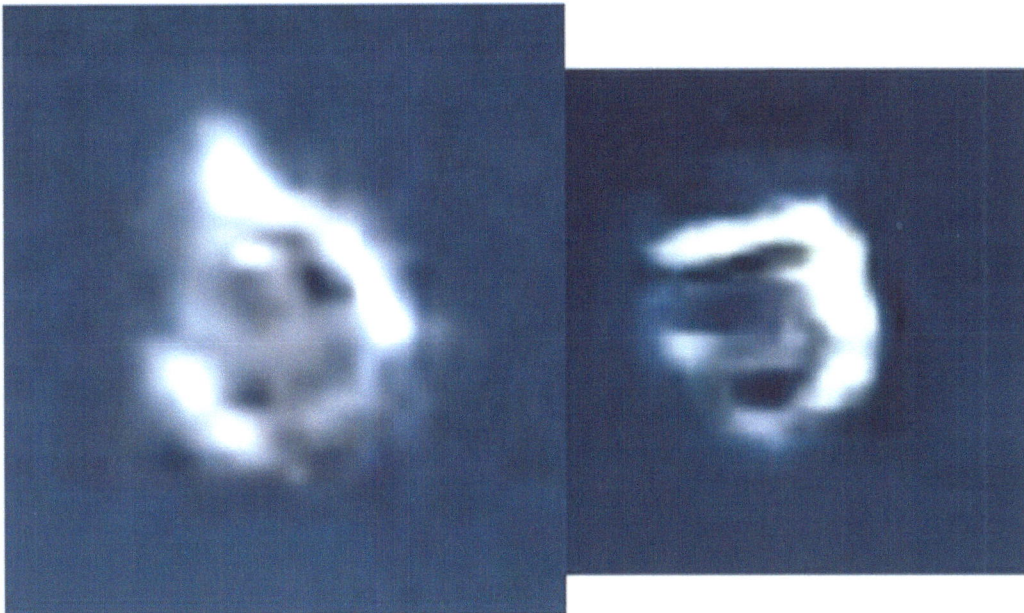

Following is part of a U.S. Patent application for a Space vehicle propelled by the pressure of inflationary vacuum state. Please visit Google Patents to see the rest of this patent and more information. This may be the type of vehicle sometimes captured on film. Patents are available for the public to view.

US Patent:US6960975B1
Space vehicle propelled by the pressure of inflationary vacuum state
Abstract

A space vehicle propelled by the pressure of inflationary vacuum state is provided comprising a hollow superconductive shield, an inner shield, a power source, a support structure, upper and lower means for generating an electromagnetic field, and a flux modulation controller. A cooled hollow superconductive shield is energized by an electromagnetic field resulting in the quantized vortices of lattice ions projecting a gravitomagnetic field that forms a spacetime curvature anomaly outside the space vehicle. The spacetime curvature imbalance, the spacetime curvature being the same as gravity, provides for the space vehicle's propulsion. The space vehicle, surrounded by the spacetime anomaly, may move at a speed approaching the lightspeed characteristic for the modified locale.

U.S. Patent   Nov. 1, 2005   Sheet 2 of 6   US 6,960,975 B1

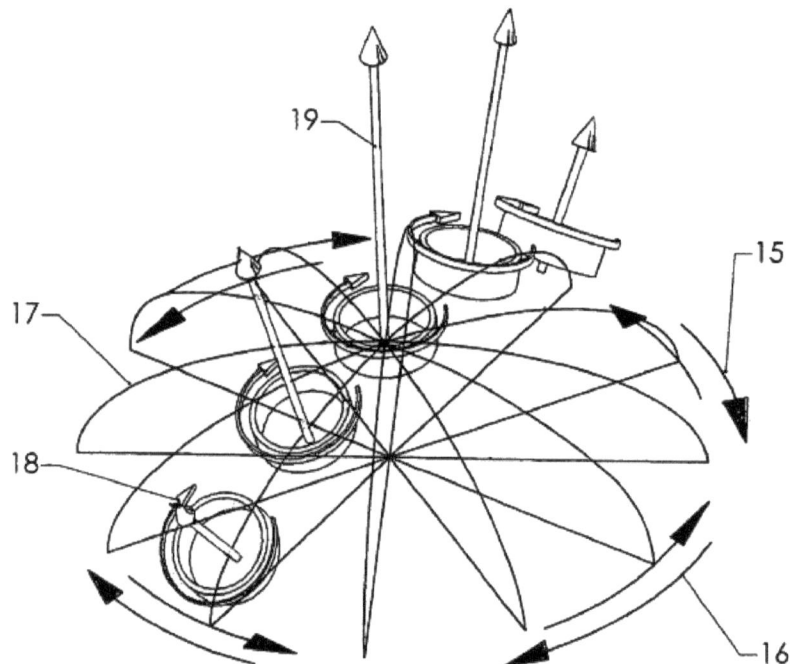

FIG. 2A

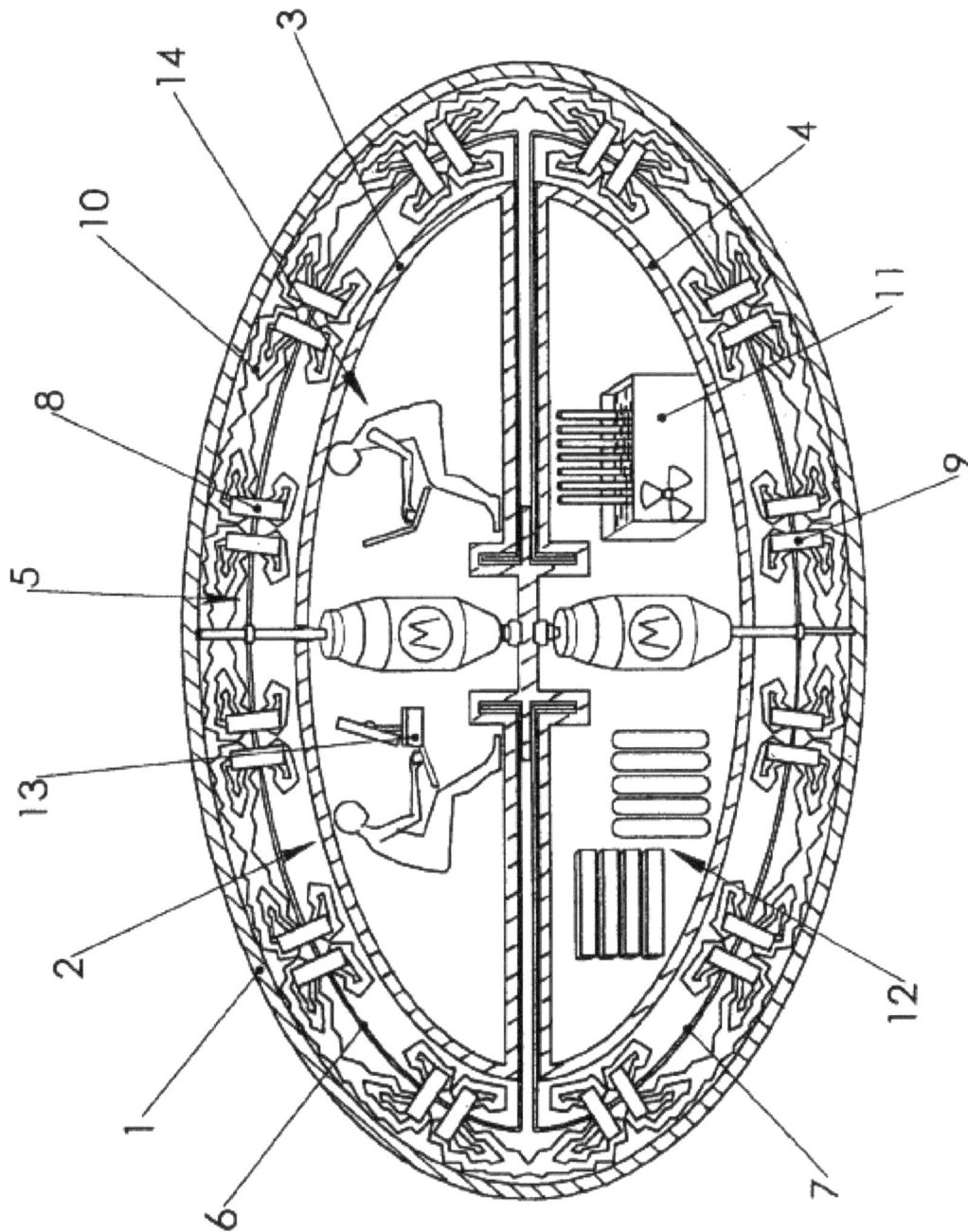

FIG. 1

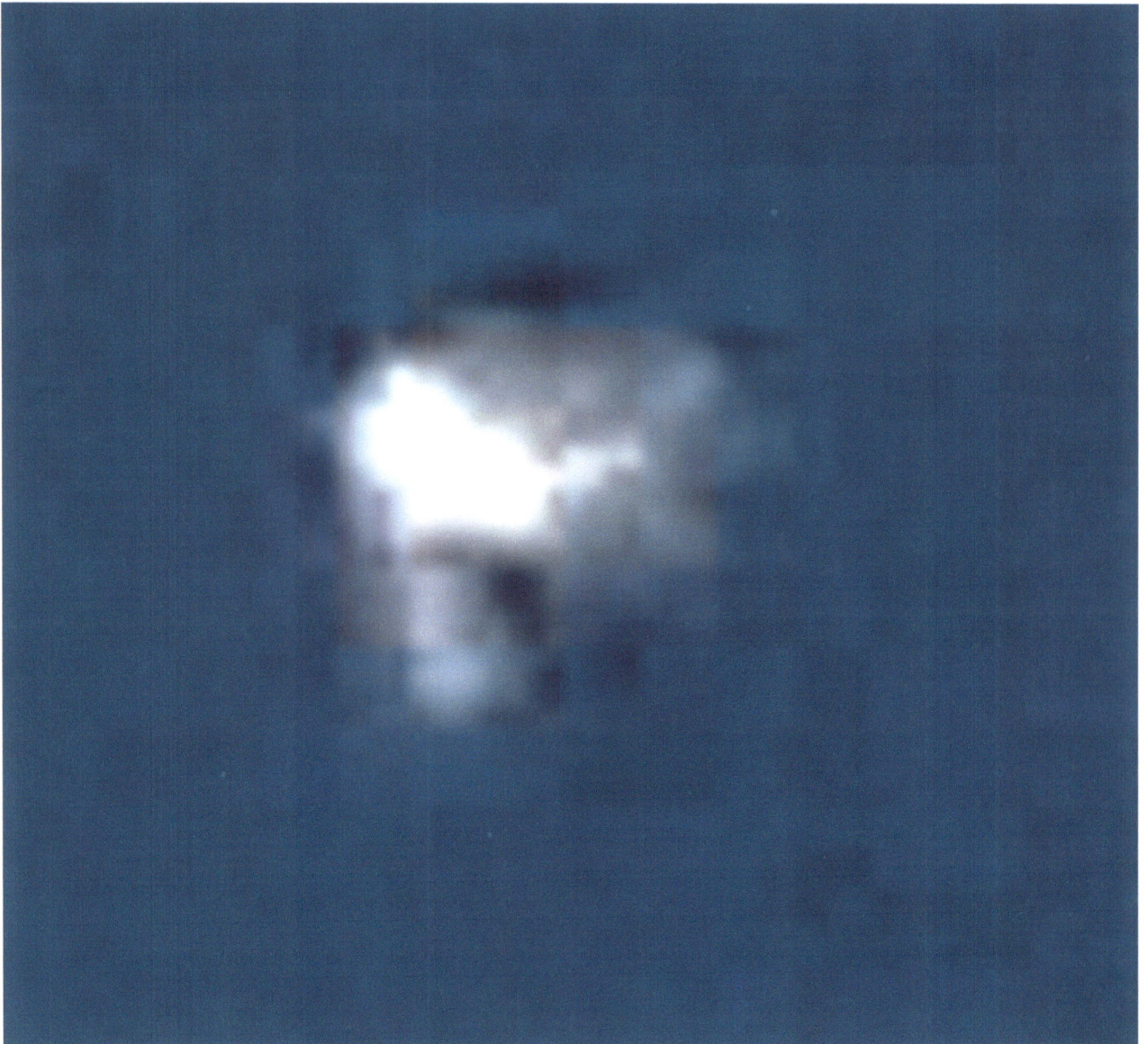

See what observations you can make, so that you may be open to experience "The Modern Miracles" yourself.

I went outside on May 5, 2020 just after 1:30 PM and took some short videos and random stills of the sky. I was in a very high consciousness state and was taking a break from working at home (Engineering Drawing). I saw a grey disk craft enter through a hole into our space/time and you can see it half in and half out in a still photo taken at full depth of field (No Zoom). I took a series of these and in several I see white orb like craft and they appeared after I asked them to. The weird part is the activity seen in my 50x power zoom when blocking the sun. There was no cloud cover, no wind and very few bugs outside at this time.

2020/05/05
01:28

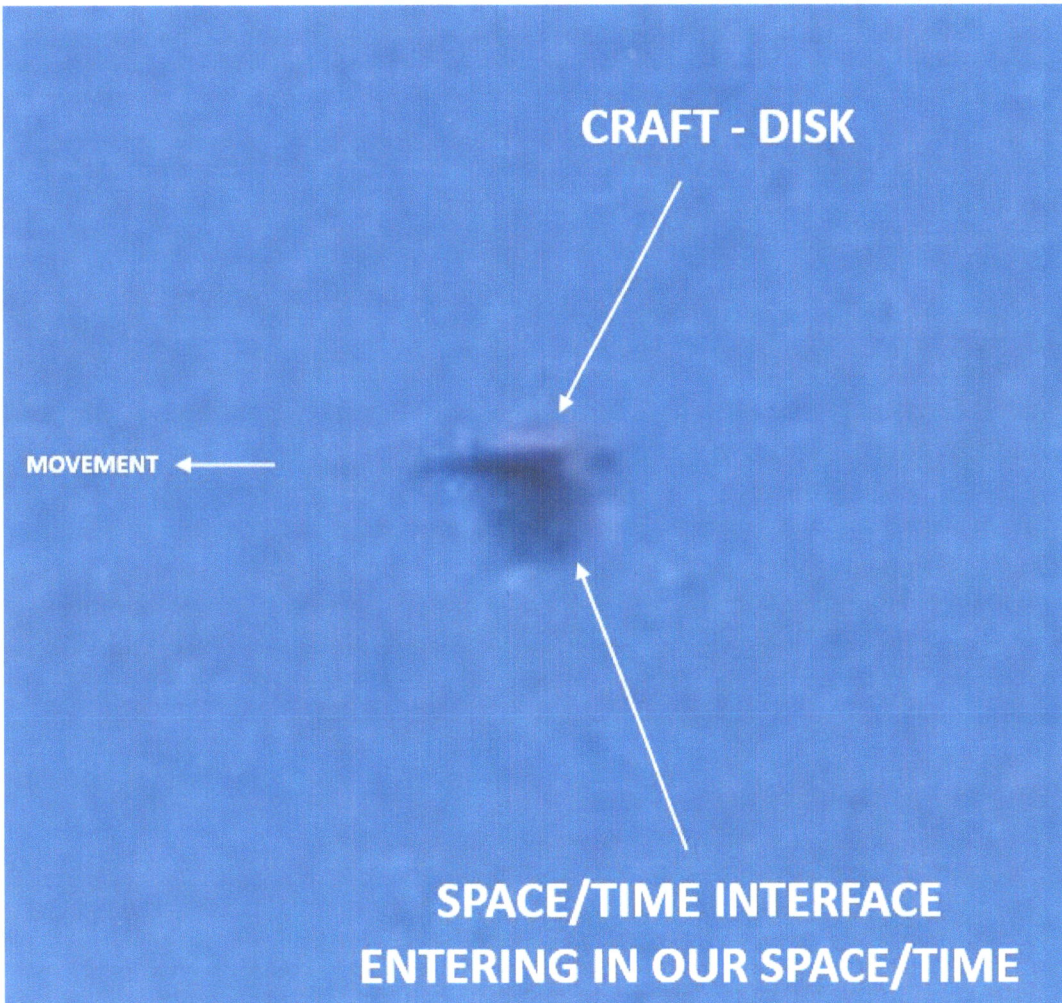

CRAFT - DISK

MOVEMENT

**SPACE/TIME INTERFACE
ENTERING IN OUR SPACE/TIME**

May 17, 2020 Precognition of UFO

Photo of craft over house. Photo: Bill Spicer

On this day just before 2:00 PM I first got the impression or precognition to go get the camera. I had been keeping track of meat on the smoker in the side yard. The whole sky near the sun was very active and I would say most of the light actually came from what I saw, not the sun. The Quantum Consciousness Field was very active. It almost over came me to even watch it through the camera view screen, since these Craft are moving so fast. It is what happened when I started to take random photos of the sky that things began to get strange.

The camera made a sound I have never heard before as I pressed the button to take still photo(s). The photos I got had a huge Craft in it and a smaller Craft similar to what I started to seeing last Summer, which is a six sided white Craft, with something in the middle forming a sacred geometry. The Wheel within a Wheel. I can not explain it, but the last two photos I took there was the largest thing I have ever seen (Cloaked) over my house.

I could see a bit of it with my own eyes. Not a cloud was in the sky. I had just asked THEM to show themselves and this happened. The camera was not in my control and took two photos. Very strange, even for me. This looked like the type of Craft I saw in my dream of what was to be my last experience while standing in a local supermarket parking lot, that I call the Flash Ascension Event. I know Margie has had her own experiences with large Craft just appearing and then go away. This was the same for me.

See the photos on the next page:

91

Enlarged photos of the object captured over the house. The first is untouched, the second is enhanced for better clarity:

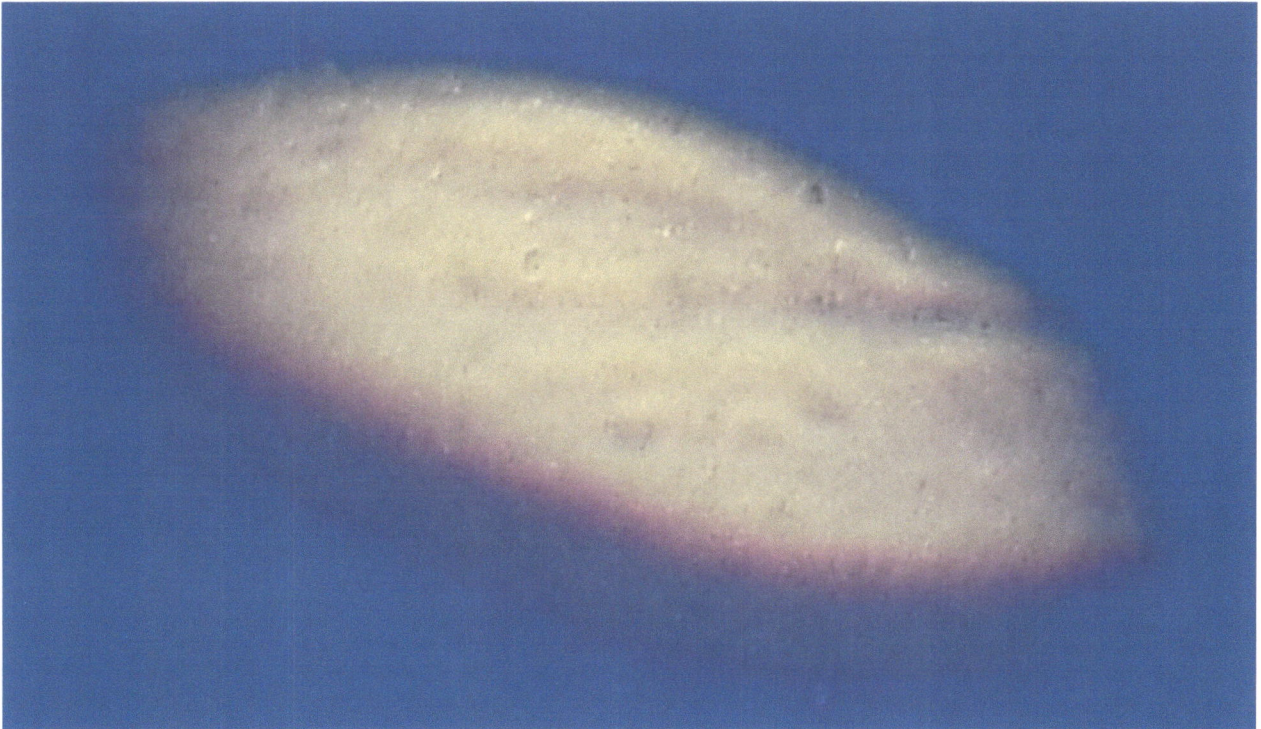

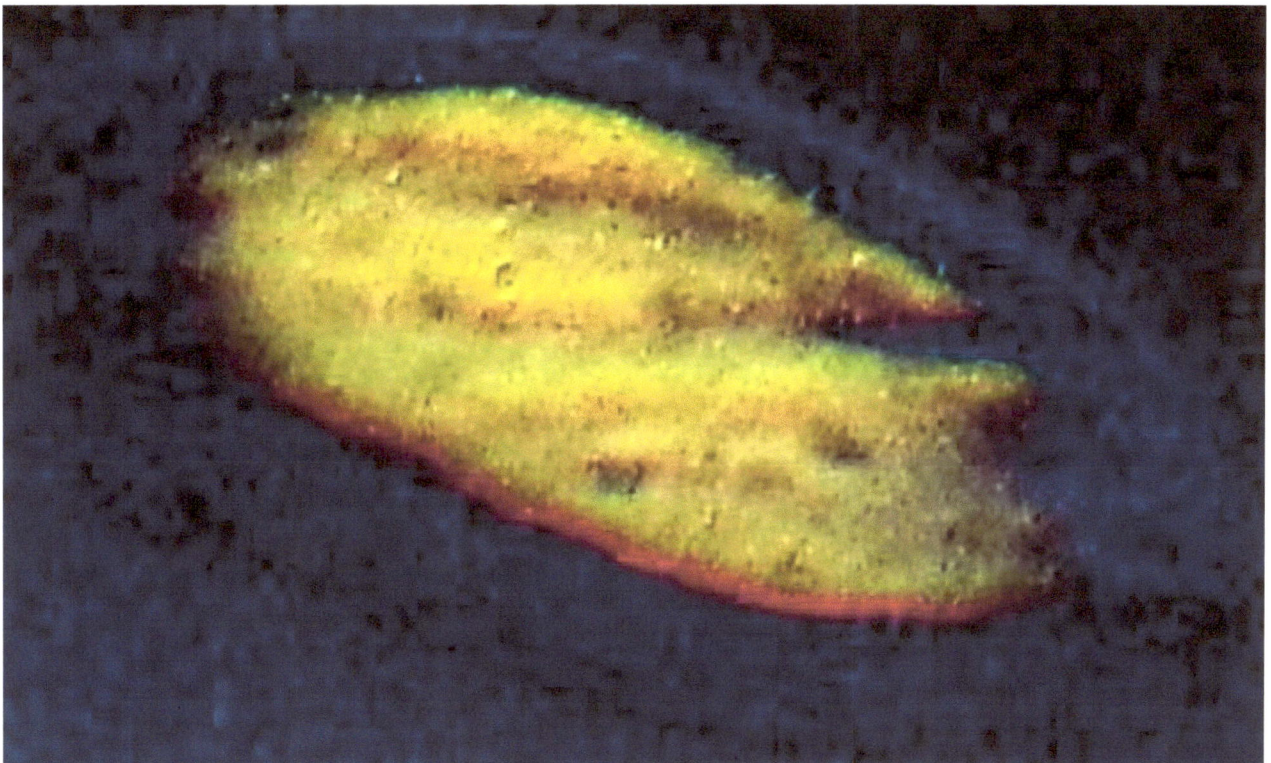

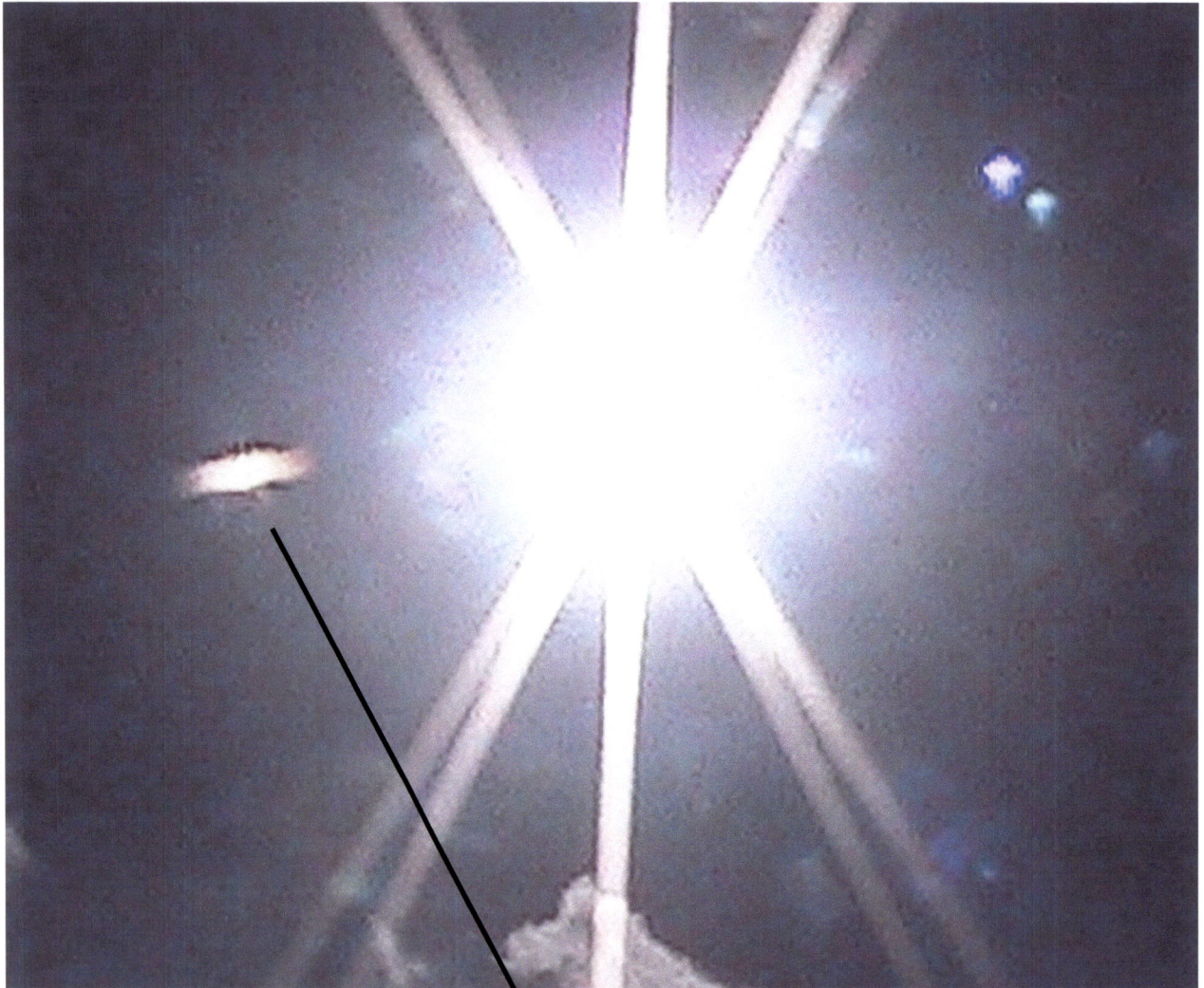

THEY ARE HERE

Craft captured by Bill Spicer using the Quantum UFO Observation Technique

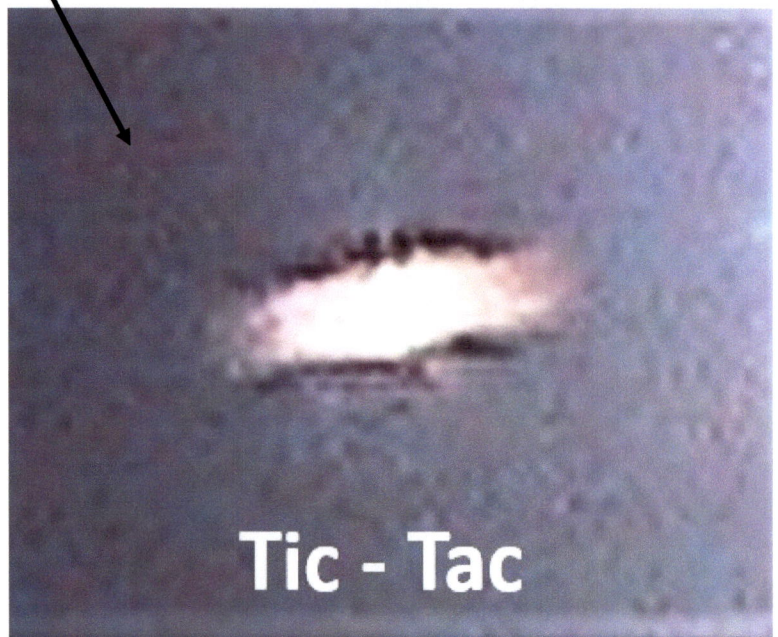

Tic - Tac

CHAPTER 3

QUANTUM TELEPATHY AND HOW TO SEE FAST MOVERS WITH THE NAKED EYE

By Margie Kay

By now, the reader knows that one method to use to see high-speed Unidentified Aerial Objects is by using a combination of mental communication and technology, and another using technology only. In this chapter, I describe my method of seeing these objects only with the naked eye. Obviously, there is no proof that one has seen something using this method alone, so if the reader wishes to obtain proof by setting up a camera, please do so.

First, some background information to help the reader understand my perspective: I became aware that I could see things others were oblivious of at an early age. By age three I realized that my parents and grandparents had no answers regarding what I later determined were energy fields around living things, and bursts of energy in the air, including negative and positive ions. By age 11 I saw my first ghostly apparition, and soon after began having telepathic communication with those who passed on, and with a guide/friend, whom I later found out is an extraterrestrial visiting Earth. Over the years my psychic abilities increased exponentially.

My parents did not encourage, nor discourage

my interest in finding out more about telepathy, ghosts, and extraterrestrials, even though neither of them had many skills in this regard. I did find out later that my mother and grandmother were very intuitive. I pursued answers at the public library and magazines, and practiced using my abilities every day for my entire life. I became very accomplished at remote viewing (see my book [1]*The Remote Viewing Workbook*), doing body scans, and communicating with inter-dimensional beings and extraterrestrials. In completing over 3,000 private readings for individuals, I've contacted thousands of spirit guides, loved ones, and extraterrestrials, and some who call themselves higher dimensional beings.

I studied the scientific side of Ufology, and became an investigator for a large UFO organization, for which I completed over 480 cases. After that many investigations, and reading about many others, I've come to the conclusion that there is only one conclusion to be had—we are being visited by beings not of the Earth. And these are highly-advanced intelligences who have likely been here for eons and long before humans.

I'm also glad to have been able to help solve

[1] The Remote Viewing Workbook describes how to activate the psychic center of the brain, which is required in order to see high-speed high-vibration UFOs and ETs. It is available at amazon.com.

over 60 homicide, theft, and missing person cases for law enforcement and private investigators and individuals using telepathy and remote viewing.

By practicing using my third eye, or psychic center of the brain on a daily basis, I've also been able to view high-speed objects that are normally not visible to the naked eye. The first fast-moving objects I noticed were not UFOs, they were the tiny bright white lights I noticed around trees, which are visible at dusk and after dark. I had been observing these tiny lights for some time when in the spring of 2006 while watching some of these lights I thought "I wonder what those are?" To my surprise, I heard a voice I knew well say "Why don't you ask them?" Although I was somewhat taken aback by this notion, I decided to send a telepathic message to the tree and its lights and was surprised to receive an immediate telepathic response:

> "We are Tree Sprites. We are the energy assigned to trees that helps them grow and know when to grow branches and leaves, and when to go dormant for the winter. We stay with a particular tree for its lifetime, then move on to a new young tree after the current one dies. We are not the same as the Tree Spirit, who is a consciousness living in the tree."

Boy, did that give me something to think about! At least now I had an answer. I watched each winter from then on as the sprites created etheric branches extending out beyond the physical branches, and sure enough, the next spring and summer the branches would grow in those exact same locations and measurements.

could see them better. Sure enough—they did so and I was able to see more of their movements. Instead of simply flitting in and out of my view, I could now see arcs like tiny shooting stars. It was only a matter of a couple of seconds, but it was long enough to see more of what the sprites were doing. This technique would be critical later on, when I wanted the Fast Movers to slow down so I could see them better.

By the summer of 2009 I came to the realization that some of the objects I saw in the skies during the daytime were actual objects and not shadows, sun flares, or optical illusions. I'd been observing these things since childhood but never paid much attention to them. There are many different sizes from very tiny to gigantic mile-wide craft. Shapes vary from tiny round pinpoints of light or dark, to snakelike shapes, saucer shapes, triangular objects, hexagonal objects, long straight lines, and others.

It is easier to see these objects in the skies right after a big rainstorm. Perhaps the charge of negative ions in the air makes it easier to see them. But the objects are also visible almost any time during the day. Daylight makes it easier to see the fast-moving objects, but in all probability, these objects are speeding about 24 hours a day.

In all cases, the objects are extremely fast moving, however, if I simply think about slowing them down, they do so, making it much easier for me to get a good look at them. I'm of the opinion that this is actually my own manipulation of time, and not the objects themselves responding. But in either case, it works.

In May of 2019 a deluge rainstorm hit Eastern Jackson County, East of Kansas City. I was

Fast-moving tree sprites observed around trees. Drawing: Kristina McPheeters

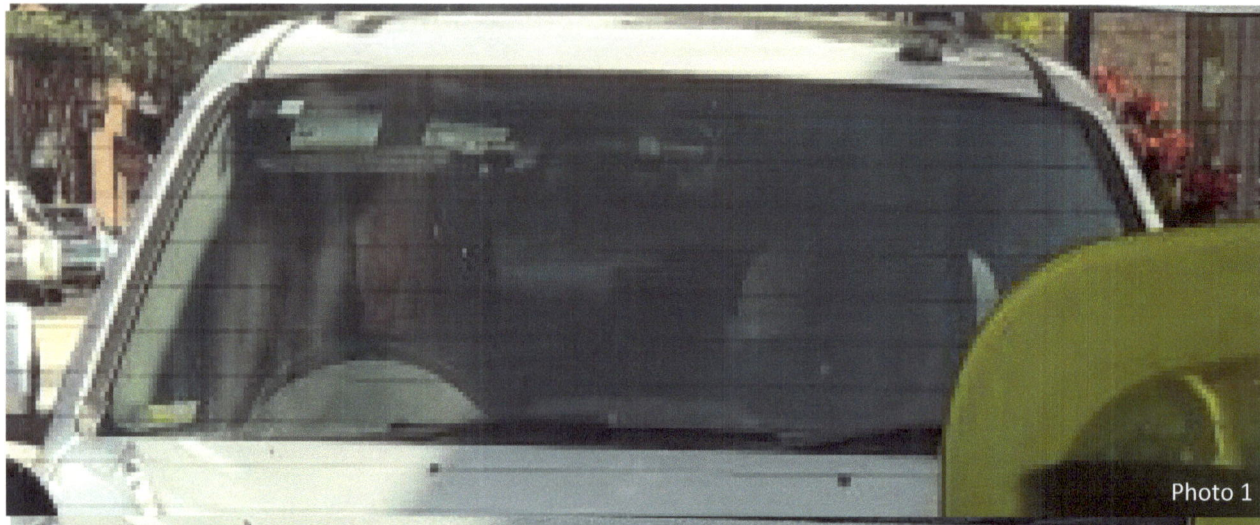

Photo 1

Photo 2

Photo 3

Slow-moving substance with strange faces in it moved across the windshield from right to left. Note the strange apparitions to the left of the driver that move in each photo, and what appears to be a bald-headed person sitting in the passenger seat. There was no one in the car with the driver. Photos: Margie Kay speaking at an event in Lee's Summit, Missouri during the rainstorm. Just before the storm hit, I was sitting in my car waiting to go inside the building where an event was held, when I noticed something very strange about the windshield of the car behind me. I watched in the rear-view mirror as a scene moved across the windshield. In this scene were several alien-looking faces, and one of the creatures was apparently sitting in the passenger side of the vehicle with the driver seemingly oblivious. I was able to gain my composure and take three photos of the windshield through my rear-view mirror.

When it was time to go inside the building, I got out of my vehicle and spoke to the man who got out of the car behind me. He was going to the same place, and there was no one in the vehicle with him. By that time I had ruled out car lights, moving reflections from vehicles, and anything that looked like what I was seeing in the area.

I came to the realization that I'd likely just observed some beings from another dimension. Perhaps the impending storm and negative ion activity had something to do with my ability to see the beings in their reflection off of the window.

After the meeting and my talk was over, I left to drive home. I immediately noticed hundreds of fast-movers in the skies and decided to call an investigator I work with often. Jean Walker is the KC Section Director for Missouri MUFON and we often work on investigations together. I knew she'd be a reliable back-up witness. As it happened, Jean was on her way home to Independence from Blue Springs, and was traveling in the same direction I was heading west. I

asked her to take a look at the skies to see if she could see anything, and sure enough, Jean saw three objects. She was the passenger in a car and was able to watch the skies unimpeded.

Although I had seen many such objects before, there were many more craft visible at that moment than normal. Unfortunately, I was driving and could not pull over to try to get photos, and frankly, the thought never crossed my mind. When witnesses see something fantastic they have a tendency to just stare at it, and not consider getting a photo or video. I did, however, draw a picture of what I saw immediately upon arriving at my office the next morning (see the next pages). Later, I asked a real artist to re-draw my awkward sketches.

The next day, I visited Jean at her house because I wanted to compare the photos I'd captured of the windshield with photos that Jean took of her house a few years ago. In Jean's pictures there are clearly different types of alien creatures, which look very similar in color and style to the creatures I captured in the windshield. Together, we decided that we must have each captured inter-dimensional creatures reflecting off of the windows. Like the high-speed craft, these entities were likely inter-dimensional. I find it fascinating that both Jean and I were able to obtain photos that look so similar.

With this in mind, I wondered if the high-speed objects in the skies were not also inter-dimensional craft that were moving so fast that most people wouldn't notice them at all. And the thought occurred to me that they could be moving in and out of worm holes or portals because in some cases, it looked like a hole opened up in the sky.

Most of the objects I saw that day were small in size. But approximately 30 grey disk-shaped UFOs appeared at the same time due west towards downtown Kansas City, Missouri. They may have been large, but that couldn't be determined without knowing the distance, which in this case would be impossible to figure. However, as I approached the intersection of 291 Highway and 23rd Street in Independence heading north, I saw the most fantastic thing I've ever seen. There in front of me, on it side, was a mile-high UFO, which appeared for approximately two seconds, but since I'd been sending a telepathic message to these objects to slow down, I was able to get a good look at it for what seemed like two to three seconds. I was astounded by what I saw, and could hardly calm down. I looked around at the other vehicles, but no one else seemed to notice the gigantic craft located in front of us.

That evening I decided to meditate and ask telepathically what the craft was doing. A voice I know well, that of [2]Valiant Thor, answered. He said that the craft was a small (!) craft sent to find some negative extraterrestrials who were attempting to hide in the Missouri River. I then realized that the Missouri River is located in the direction I was looking.

After checking on Google Earth, I found that if the UFO was indeed over the Missouri River, it was six miles from my location. That meant that the craft was huge.

Its perplexing to think that hundreds, or even thousands of craft could be entering and exiting our dimension every minute of every day. If this is indeed the case, that could change our perspective of UFO's altogether.

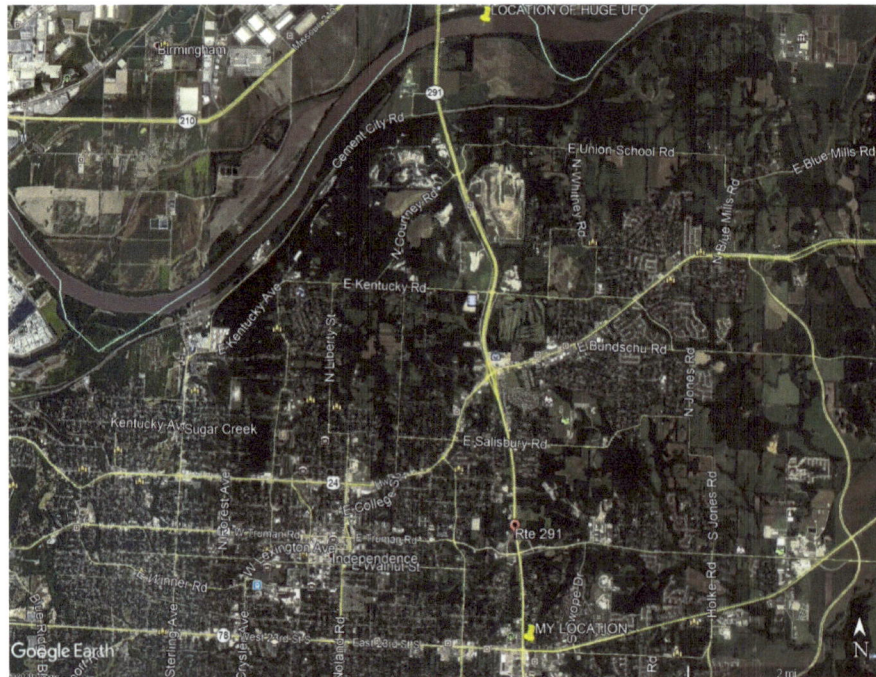

Map showing my location and the assumed location of the UFO, which is 6 miles distant.

(2)Valiant Thor is an extraterrestrial who visited the Pentagon from 1957 to 1961. He contacted me in 1985 and has been in contact with me since. See more information in my upcoming book THOR, and in the book Stranger at the Pentagon by Frank Stranges. Thor is the commander of a large fleet assigned to protect the Earth and advancer humanity.

Fast Movers over downtown Kansas City in May of 2019 Drawing: Kristina McPheeters

Large UFO on its side observed by Margie Kay from the intersection of 291 Highway and 23rd Street in Independence, Missouri in May of 2019. This was only observed for approximately two seconds.

Drawing: Kristina McPheeters

How to See Fast Movers with the Naked Eye

I believe that everyone can learn to see high-speed craft, and in fact, anything from other dimensions or frequencies because we all have the same physiology—i.e. the psychic or PSI center of the brain, which is responsible for the so-called Sixth Sense.

By activating this center, which includes the pineal gland central to telepathic function, a person can learn to see things that are at a higher vibration than normal.

The most important tip I can give anyone is to **practice meditation daily**. By doing this, a person will activate the pineal gland, and trigger an awakening of the PSI center which is responsible for telepathy, clairvoyance, clairaudience, and clairsentience. Not to mention the other more advanced senses including Remote Viewing.

In order to see the Fast Movers, a person must first become aware of their existence. Next, they must want to see them. And finally, they must attempt to see them.

The beings piloting these craft are higher-frequency entities. Most are benevolent beings with only positive intentions. And many are willing to, and enjoy working with humans who are waking up and are happy to assist us in this endeavor. It is for this reason that I believe they slow down for me when I ask, or perhaps I put myself into another time outside of normal time. I have yet to determine which method it

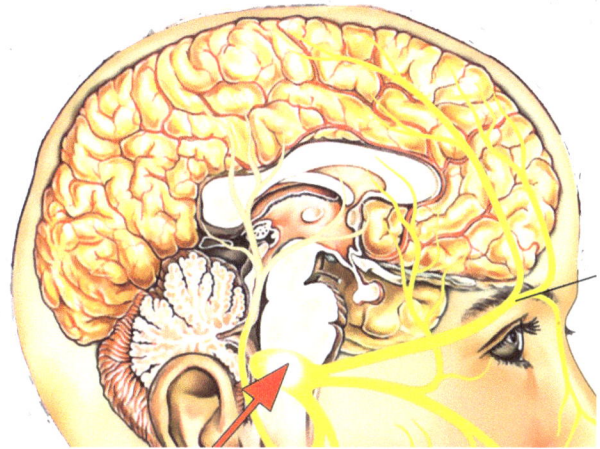

The PSI center of the brain

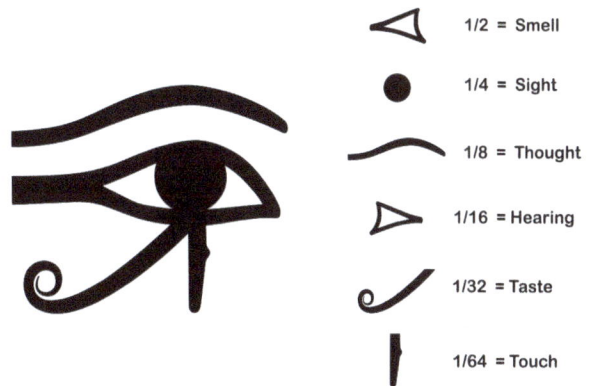

The Eye of Horus represents the PSI center

actually is, but it matters little as long as it works.

Meditation is the first step to viewing the Fast Movers. I suggest going to a meditation training center if at all possible, or take a course online. Following is a meditation method that I use and most people have success with it:

102

Margie's Meditation Method to Access the PSI Center of the Brain

Meditation can help you get to the Beta level of consciousness, where psychic abilities are more easily accessed. At the Beta level our brain waves average at around 12—15 cycles per second. At Theta state brainwaves are between 6-7 HZ , and at Alpha (normal waking and relaxed state) brainwaves are between 7 and 12 HZ. Beta or Theta level is the state we try to achieve before contacting inter-dimensional beings of a higher frequency.

1.) Pick a time to practice every day, preferably at night when it is quiet and there is little activity going on. Try to stick to the same time every day. I meditate just before going to bed.

2.) Pick a place to meditate that is comfortable and where you won't be disturbed by people, pets, or loud noises like traffic or trains.

3.) Either sit in an easy chair, or lie down on a bed or couch and use pillows or blankets to get very comfortable. I prefer to lie down because it is more comfortable for me. If you fall asleep while meditating don't worry about it. *Note: Meditation before sleep often allows you to have a deeper, more restful sleep.*

4.) Use of a meditation CD is a good way to get the brain down to beta state fairly quickly. Listen to the music and relax. If you don't have a CD, just listen to your own breathing for a few minutes. I can highly recommend Janalea Hoffman's CD's (see the resource listing)

5.) Get relaxed: start at your toes and flex, then relax your muscles from your feet to the top of your head. Start with the feet muscles, then calves, then thighs, etc. all the way to your neck and head. Go back a couple of times to the jaw and neck areas, since this is the area where people hold the most tension. Take your time and don't rush the process. After using this method for a few months, you'll get very adept at relaxing and won't have to go through the entire process anymore. Your entire body will relax at your suggestion.

6.) Clear your mind. First visualize a blank black curtain in front of you with nothing in front of it. You can do any number of activities at this point, based on what your goals are. If you want to contact a loved one who has passed on, or a benevolent being of any type, ask to be in contact with that person. Visualize them standing in front of you. When you have a clear picture, ask the person a question and listen for a response. If you don't succeed at first, keep trying. After several sessions you should be able to hear or "feel" the answer. Be sure to write down everything that happened after you stop meditating.

Another project you may want to do while meditating is to create something. This will also help you get used to using the subconscious mind. Then visualize whatever it is you would like to manifest in your life—a loved one who passed on that you would like to communicate with, or money, love, a house, job, travel, health, or whatever you wish. Concentrate on that and say to yourself "This manifests for me now" or something similar, over and over again until you feel confident that you have created this in your life. Believe that this is happening, and it will happen.

Always use positive words instead of negative words. For instance, say "I am completely healthy" instead of "I don't have arthritis," since the subconscious mind understands only what you are concentrating on and visualizing.

Start with small things that are easier to obtain before moving on to bigger, more important things. You'll be surprised how your life can change using meditation on a daily basis. Don't be discouraged if you have trouble relaxing and concentrating at first. A lot of people find it very difficult to stop for 15-30 minutes a day after a hectic day.

Your mind has to get used to the change, and it may take a while— but after a few weeks you'll start to feel more confident and it will get easier and easier. It does not matter what the focus is during your meditation practice - the process will train your mind for ET contact.

I have been meditating for 45 years, and I strongly believe that it has helped me train my mind to the point that I can instantly go into a trance state anywhere at any time. Some people get to the stage where they can have "Dual Consciousness," which means that part of your mind is in a trance state, and part of your mind is fully awake, alert, and communicating with others around you.

After working on meditating for a while, try your hand at seeing the Tree Sprites. Go outside on a clear night with no lights nearby, sit in a chair, and look at the top of a tree. Relax your face and eyes. Only by relaxing your eyes will your night vision kick in. Don't stare hard or it won't work. It helps if you are a bit tired.

After a few minutes, you should see a thin dark outline around the leaves, then a lighter thicker light line. This is the energy field of the tree. Next, you should see tiny bright white lights. At first, you'll only see these with your peripheral vision, and that's OK. After doing this for a few weeks or months, you will finally be able to look directly at the sprites and see them.

Now ask them to slow down so you can get a good look. Simply say: "Could you please slow down so I can see you better?" Whether the sprites do this or it is you slowing time down, it will likely work. The sprites may even communicate with you telepathically. Ask them *questions and see what response you get.*

If you can see the Tree Sprites, you can see the Fast Movers!

Best Time to See Fast Movers:

I see the high-speed UFOs best during the day against a cloudy sky, but they are also visible on clear days. They seem to be the most noticeable while I'm driving to and from work ore going on a trip. While driving or doing any repetitive activity, a sort of mild road hypnosis can occur to anyone, and this is the state of mind a person needs to be in to see the most activity. However, I don't suggest listening to an hypnosis CD while driving. The normal routine of driving will do it.

A person can also see these objects by simply sitting outside and relaxing for a while. Pick a quiet place with out disturbances. Perhaps read a book or magazine. Then, after you are fully relaxed, start scanning the skies. The Fast Movers will appear as short blips in many different shapes and sizes. You may mistake them for optical illusions, eye floaters, or a smudge on your glasses at first, but soon you'll begin to see many of these objects. Watch for light and dark craft. It seems that they move too fast to see other colors or lights. I've seen bright white, dark grey and black, and metallic grey objects. Many of the

pinpoints of light are so bright they are blinding, You may also see bright white flashes of light night and day. I have come to the conclusion that these are likely Fast-Movers moving through a temporary inter-dimensional portal opening.

Fast-Mover Shapes to Look for:

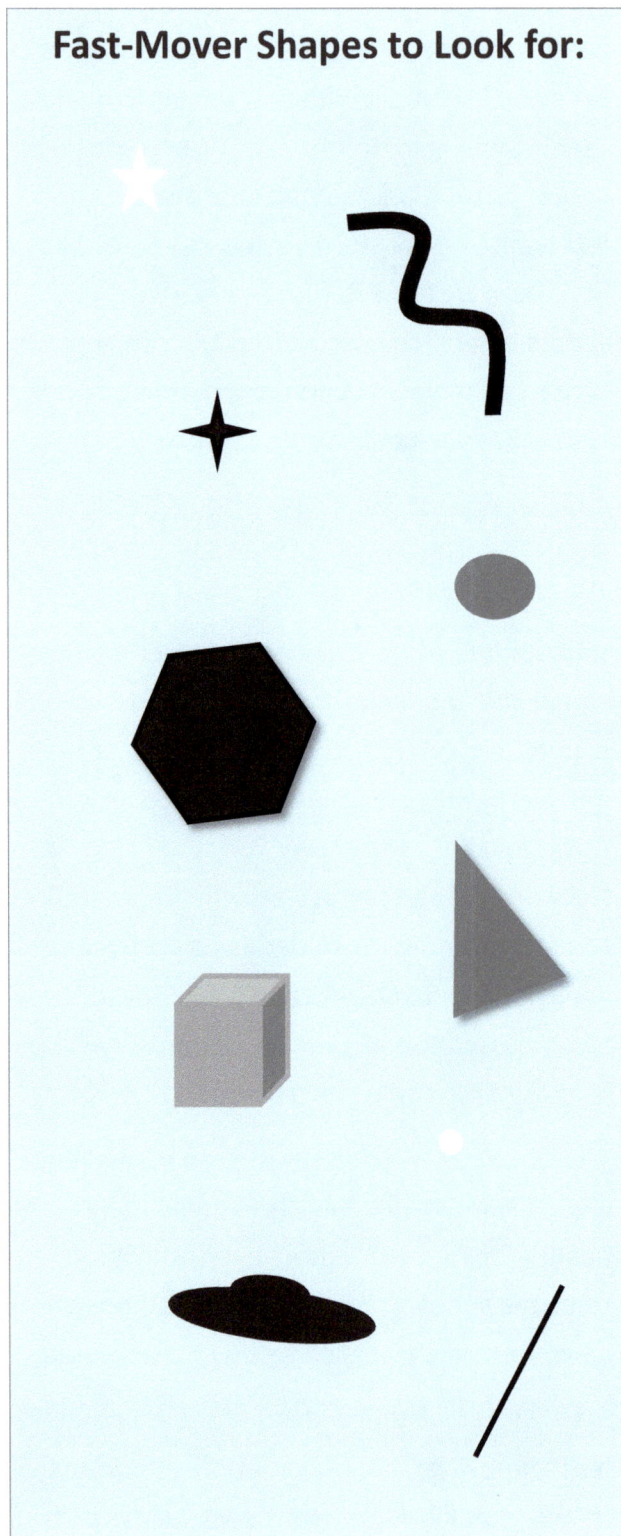

By Margie Kay, Wayne Lawrence, and Bill Spicer

Blinding Flash During an Investigation:

During an investigation in Henry County, Missouri which was covered on TV news shows and in a *Hangar 1* episode, something strange happened that I now believe had to do with a Fast-Mover.

I investigated a cow mutilation on a ranch in July of 2013. The owner had the feeling she was being watched the day before this event occurred, and actually took photos of the sky, even though she could see nothing. The next day a cow was found mutilated and dead. A few days later, a second cow was found in the same condition as the first.

During the second trip to this location I was standing near the site where the cow had been found when suddenly there was a bright flash of light which lit up a 3' diameter circular area on the ground less than three feet from where I was standing. The flash was so bright that it blinded me for a few seconds. The witness and another investigator were facing in another direction and did not see it. I have been perplexed about this event for some time, but now realize that it was likely a craft moving through a portal, which opened for just a second. Could the portal have been on the ground? Perhaps the answer is yes.

The proximity to the location of the dead cow (it had been removed, leaving a strange burned outline on the ground) cannot be coincidental. I experienced other strange events during my visits to this site including audible planes going overhead at low altitude that were not visible, and a sudden gush of wind in a dead calm day

Site of animal mutilation event in Henry County, MO and location of circular light on the ground.
Photo: Margie Kay

that made us think of the wind produced when a helicopter starts up. Something strange occurred at this site, and if Fast-Movers were involved it wouldn't surprise me. Perhaps this could also explain crop circles that appear in just a few minutes, and many other strange phenomena that perplex investigators.

The Fast-Moving Entities

After observing the tree sprites and craft in our skies, I realized something. Since the craft are high-speed UFOs, what if extraterrestrials and other types of non-human entities are also fast-movers? This would make sense considering all of the people who see shadows, shapes, and lights moving about inside their homes. In many instances, people say that they see something in the peripheral vision that moved very fast. They know something is there, but then could not focus on it.

I belong to a group who communicates via email on a daily basis. All of these people are credible, sane, and intelligent observers. Many of them have seen fast-moving beings in their houses. I'm sure that what they, and I, see on a daily basis are things that are really there.

I had a strange event occur during a UFO investigation a few years ago in which an ET appeared three times, each time only for a second or two. After pondering this for some time I've come to the conclusion that the ET was standing in the same spot the entire time, and I only perceived him at those particular instances due to my own state of mind (and being tired) at the moment. I had changed my state of consciousness to theta level, and this is why I was the only one out of the six persons at the site who saw the entity. He was a higher-vibration being, or Fast-Mover if you will.

Why do we perceive these entities and craft as Fast-Movers? Perhaps it is because we are in a lower and slower vibration third/fourth dimension, and they, occupying the same space, are in the fifth and higher dimensions which vibrate at a higher rate. To us they are very fast, while to them, it may not be fast at all.

UFO Propulsion Systems

Several years ago I telepathically contacted Valiant Thor and requested a visit on his craft which is stationed on Earth. I was able to do this using my etheric body during a Remote Viewing session. The engineer met with me and took me to the central part of the craft, and we moved into a very large round area with a railing around it. Looking down and up I could see outside the craft. The engineer ex-

plained that they used two propulsion systems, one for high-speed which she called an Ion anti-gravity photon system. She also said that light photons were a by-product of sorts when the system was turned on, and this is why their craft are literally covered in light when they are using this system. The reader may wish to read an article I wrote about this on my blog at www.margiekay.com. My point in mentioning this is that Bill Spicer and I independently received similar information from our higher-dimension contacts.

The Hummingbird Effect:

Hummingbirds flap their wings at up to 200 times per second, making it nearly impossible to see the wings with the naked eye. The birds also flit about very quickly, almost appearing to be in two locations at the same time. For us, to consider moving that fast would be impossible, yet it is normal for hummingbirds. The photo on page 113 could only be captured by using a high-speed camera. Otherwise, the wings would simply be a blur to the observer. This is analogous to the Fast Movers. Perhaps hummingbirds are physiologically at a higher vibration than most creatures on earth.

This begs the question: Just how many fast-moving higher-vibration beings exist? It is entirely possible that there are millions of beings that exist in our multiverse which we are only just now starting to perceive with our own senses and with the assistance of technology.

If we consider the electromagnetic spectrum, for instance, this shows the range of waves that we know to exist.

A hummingbird captured in mid-flight with a high-speed camera

Photo: Adobe Stock

Gamma rays
10^{-12}

X-rays
10^{-10}

Ultraviolet
10^{-8}

Visible light
0.5×10^{-6}

Infrared
10^{-5}

Microwave
10^{-2}

Radio
10^{3}

It may be that the Fast Movers operate on a high vibratory rate that is just outside most human's ability to not only perceive them, but interact with them.

Wave Lengths

Notice how the waves or vibratory rate is different at each level.

Psychic Photography and Interdimensionals

Psychic Photography was discovered in 1862 by William H. Mumler, an engraver in Boston, Massachusetts. Mumler had been experimenting with the new invention of the camera when he inexplicably captured spirits along with is family members in photographs. News of this discovery soon spread around the world and others began their own experiments.

F.W. Warrick, a British parapsychologist published *Experiments in Psychics*, in which he

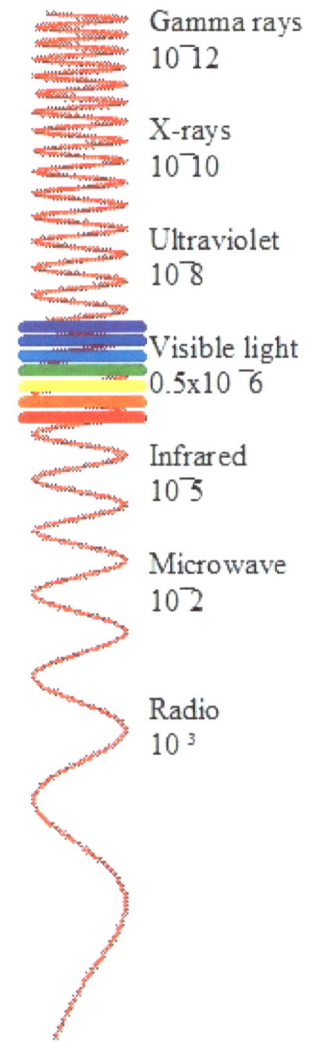

displays hundreds of photographs of spirits while a psychic medium attempted contact with the other side.

I suggest reading the book *Psychic Photography: The Visual Proof* by Hans Holzer, PhD for more information and photographic evidence of spirits, which in my opinion exist in the higher dimensions along with the Fast Movers.

I discovered psychic photography while working on a combined UFO and paranormal investigation in a small town in Missouri where many

THE ELECTROMAGNETIC SPECTRUM

Wavelength (in meters)	10^3	10^2	10^1	1	10^{-1}	10^{-2}	10^{-3}	10^{-4}	10^{-5}	10^{-6}	10^{-7}	10^{-8}	10^{-9}	10^{-10}	10^{-11}	10^{-12}

longer → shorter

Size of a wavelength: Soccer Field, House, Baseball, This Period, Cell, Bacteria, Virus, Protein, Water Molecule

Common name of wave: RADIO WAVES, MICROWAVES, INFRARED, VISIBLE, ULTRAVIOLET, "SOFT" X RAYS, "HARD" X RAYS, GAMMA RAYS

Sources: AM Radio, rf Cavity, FM Radio, Microwave Oven, Radar, People, Light Bulb, The ALS, X-Ray Machines, Radioactive Elements

Frequency (waves per second)	10^6	10^7	10^8	10^9	10^{10}	10^{11}	10^{12}	10^{13}	10^{14}	10^{15}	10^{16}	10^{17}	10^{18}	10^{19}	10^{20}

lower → higher

Energy of one photon (electron volts)	10^{-9}	10^{-8}	10^{-7}	10^{-6}	10^{-5}	10^{-4}	10^{-3}	10^{-2}	10^{-1}	1	10^1	10^2	10^3	10^4	10^5	10^6

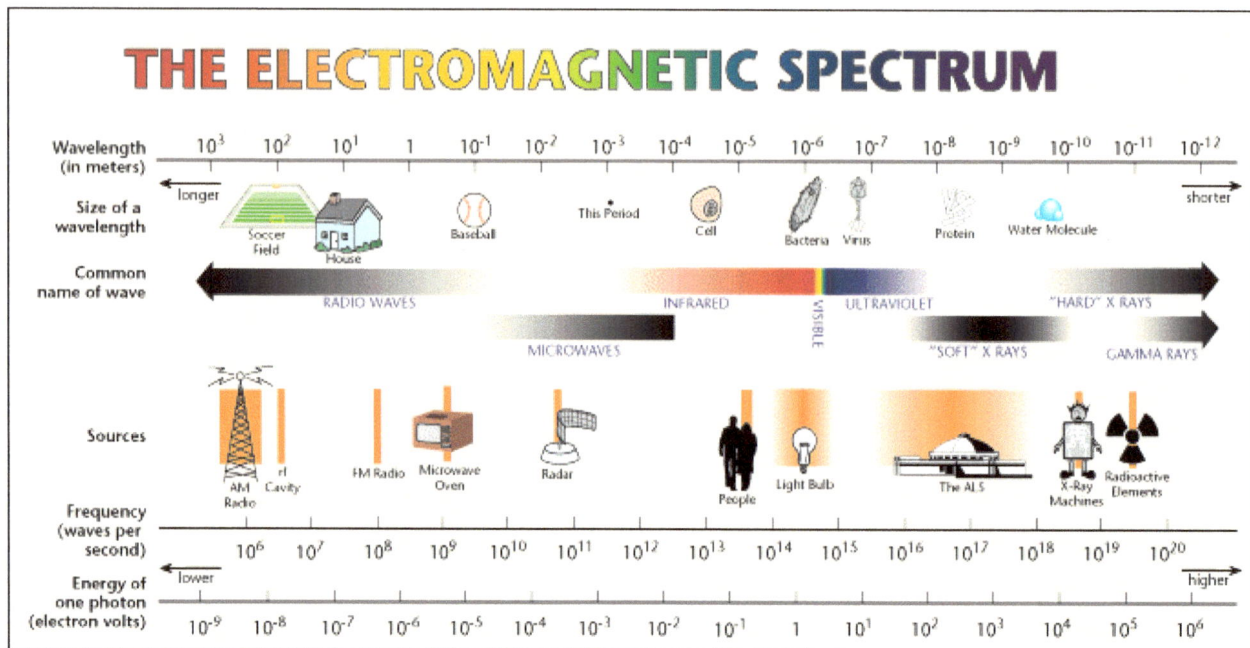

The visible light spectrum is a very small part of the known electromagnetic spectrum. Perhaps our visitors known as the Fast Movers, are normally just outside our visual perception.

unexplained things occur. The primary witness tried an experiment one evening after he had been tormented by objects moving about his house on their own. The case is very complicated and involves flying craft chasing the witness' vehicle and in two instances, crashing down on top of the vehicle leaving a huge indentation; possible abductions; photos of extraterrestrial; and more. The witness practiced meditation, and wanted answers so he thought he would try psychic photography, although I don't believe he knew what the method was at the time. He just had an idea to try it.

The witness first went into a trance state, then raised his Cannon digital camera and took a photo of a side table against a wall in his living room where objects seemed to disappear, then reappear in a different location in the house. He wondered if he could capture anything and was shocked to see a gold colored round metallic object with six ports and red lights on it, similar to a tiny UFO in one photo. He immediately sent me the picture via email.

According to the time stamp on the photo, it had been taken only a couple of minutes before I received it in my inbox, therefore, it couldn't have been manipulated. Here was a physical-looking object in a photograph which could not be seen or felt in the third dimension. It may have actually been a UFO, as I've worked on a number of cases involving tiny unidentified flying objects. I told the witness to keep taking pictures in different rooms in the house in order to help unravel the mystery of this house, which was a hotbed of paranormal activity.

He continued to do so, and began to turn the camera around and take a picture of himself while meditating and in so doing captured some amazing pictures of lines of light energy and even alien-looking creatures standing next to him. Unfortunately, I can't share these pictures because he won't give permission to do so.

As it happened, the home was in the direct path of a Leyline, which may also have had something to do with the amount of activity at this particular location.

Seeing Inter-Dimensional Beings with the Naked Eye

Yes, you read correctly. Most people who are able to see UFOs/UAPs, whether they are Fast Movers or Slow Movers, are also able to see Extraterrestrial beings. It is the next step in the evolution of consciousness. The fact is that most people are too afraid to try to see them. I completely understand this as these other beings don't look like humans, and it is a shock to see one the first few times.

My First In-Person Experience

A few years ago I sent a telepathic message out to any benevolent beings to assist me with a severe problem I was having with my feet that the doctors could not diagnose. I then fell asleep but was awakened by a sharp pain in my left calf like a needle going in to the bone. When I opened my eyes I saw two short, thin extraterrestrials standing at the foot of my bed. One had his hands over my feet. Shocked at the sight, I screamed and they disappeared. After I recovered from the shock, I realized that they were still there, but I had lowered my vibration and consciousness so could not see them any longer. The next day, after weeks of pain, I no longer had pain in my feet. They answered by request for help, but I reacted in fear.

It took four visits from extraterrestrial/

interdimensional beings for me to get comfortable with the idea of seeing them with the naked eye and communicating with one or several while looking at them. It is one thing to have telepathic contact, and yet another when they are standing in front of you.

One thing to note is that all of these beings may not be Extra-terrestrial at all, but rather inter-dimensional and sharing our same space.

What I have observed is that these beings vibrate at a high rate, and are moving about in the fifth and higher dimensions. They are Fast-Movers, if you will. When a human has reached a certain level of consciousnesses, thereby raising his vibratory rate he/she will be able to see these high-vibration beings.

Since my first experience I 've been in contact with a number of different types of non-human entities who have assisted me with medical issues, answered many questions, and have even done some rather humorous things. In all cases, these entities communicate telepathically and are far more advanced than humans.

What Thor says about our observations:

During a meditation on 8/27/2020 I was given further information about Fast Movers by Valiant Thor. He said that humans perceive high-vibration beings and craft as fast moving, but the beings themselves are moving at what they consider to be a normal pace. Therefore, they have no ill effects when their craft move at 90-degree angles or what we perceive to be high velocity. They are actually in a different space/time than we are. As we evolve, humans will eventually perceive their craft to be moving at what we consider to be normal speeds because our vibratory rate has increased.

He also stated that as we increase our level of consciousness and vibration we will eventually be in the fifth dimension at all times, and people who remain in the third dimension will no longer be able to see or interact with us. He also stated that is what happened to the lost civilizations such as the Maya and others. They made a conscious decision to shift permanently. Some did decide to stay behind.

Thor then said that all dimensions exist in the same space and that the same laws of physics apply to each dimension. As each individual increases their vibration, they have the choice at some point to move to the next. Perhaps this is what is called "Ascension."

Conclusion:

Take some cases of human interaction with perceived extraterrestrial, or perhaps inter-dimensional beings: In many accounts the observer sees one of these entities for just a second, then they disappear, or in some cases they reappear almost immediately at another place. Did the entities raise their vibratory rate so fast than humans can't perceive them? Or in some cases, humans do interact with these beings for long period of time. Does this mean that the being lowered its vibratory rate, or that the human raised theirs?

I lean towards the second theory—humans are raising their level of consciousness and vibratory rate, and are becoming more aware of what has been around them all along. Just look at all of the documented videos that have

surfaced as of late and the TV shows that cover these phenomena. Due to the collective increase in consciousness and awareness across the globe, the Fast-movers are becoming more and more visible, and the person holding the camera is the key in most cases.

These are questions that need answers, and I believe that we are on the verge of getting them.

We are just getting to the point of understanding the Fast-Mover phenomena. Thanks to the work by Bill and Wayne, I have a better grasp of the situation and realize that what I've been observing in the skies for years is now validated by their work in capturing images on film, or by analysis of footage taken by others. Thanks to these gentlemen, we now have clear evidence of the phenomenon.

Those of us with psychic abilities and those with scientific backgrounds should work together to gain a better understanding of the Fast Movers and their occupants.

As Nicola Tessla said, "If you want the answers to the Universe, think in terms of Energy, Frequency, and Vibration." In my opinion, our ability to observe and interact with these high-frequency craft and beings is entirely dependent on our own vibratory rate and state of consciousness.

Tessla was right.

Chapter 4
Fast-Mover Images Captured by Others

Kerry Walker, host of a YouTube talk show from Houston, Texas, used the Quantum UFO Observation Technique to capture images of Fast Movers. She was introduced to the method by Bill Spicer in April of 2019 and tried it out using her iPhone 8 with a polarized lens which was removed from an inexpensive pair of sunglasses and taped to the camera.

Kerry has taken videos using her Nikon camera as well, but prefers the iPhone since she always has it with her.

Kerry says she is careful to watch out for bugs and doesn't want to mistake them for something else.

On November 9, 2019, Kerry was able to capture three different objects over her house (or perhaps the same object three times). At right are the original photos with enlargements included.

Kerry Walker
Kerry@AliceEatsTheApple.com
www.AliceEatsTheApple.com

More photos from Kerry:

Original photo at left, enlargement above right. Kerry calls these objects "sticks".

Below: Photo of object captured using the QUFO Technique.

Kerry took this photo in San Miguel Mexico. She did not see the object in the sky with the naked eye a the time, but found it later while reviewing her pictures.

Perhaps this was a Fast-Mover which was captured at just the right moment (or 30th of a second).

It is not uncommon for people to find UAPs in photos which they did not observe at the time the picture was taken.

Below: The object enhanced and in black and white to see more detail:

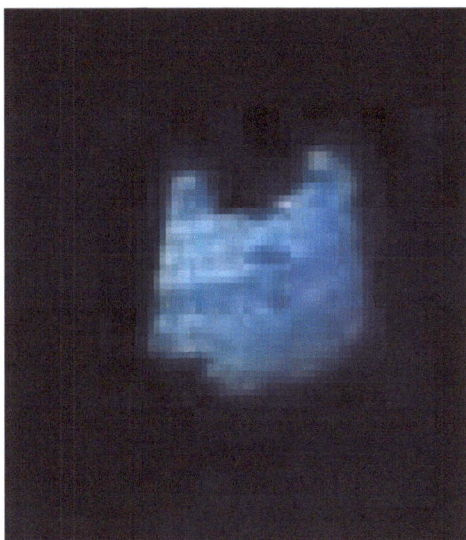

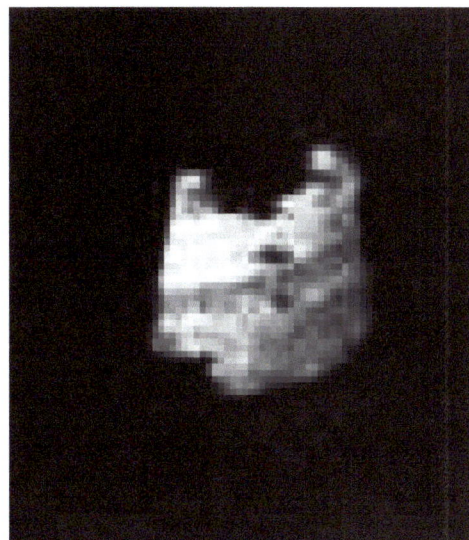

A Winged Creature Captured on Film

Tony Degn was walking on a street in Lee's Summit, Missouri in 2014 when he noticed a black helicopter flying nearby. It had no markings on it. Curious, Tony snapped a couple of pictures of the helicopter using his cell phone, then when he reviewed the pictures later he noticed something else in the photos that at the time was not visible to the naked eye. Tony believes that there is a humanoid winged creature with legs flying next to the helicopter. He had seen a creature similar to this one shortly before this incident, but that time he had no camera and saw the winged man with the naked eye. Could this creature have been a fast-mover?

My study of these winged humanoids leads me to believe that they are actually aliens.

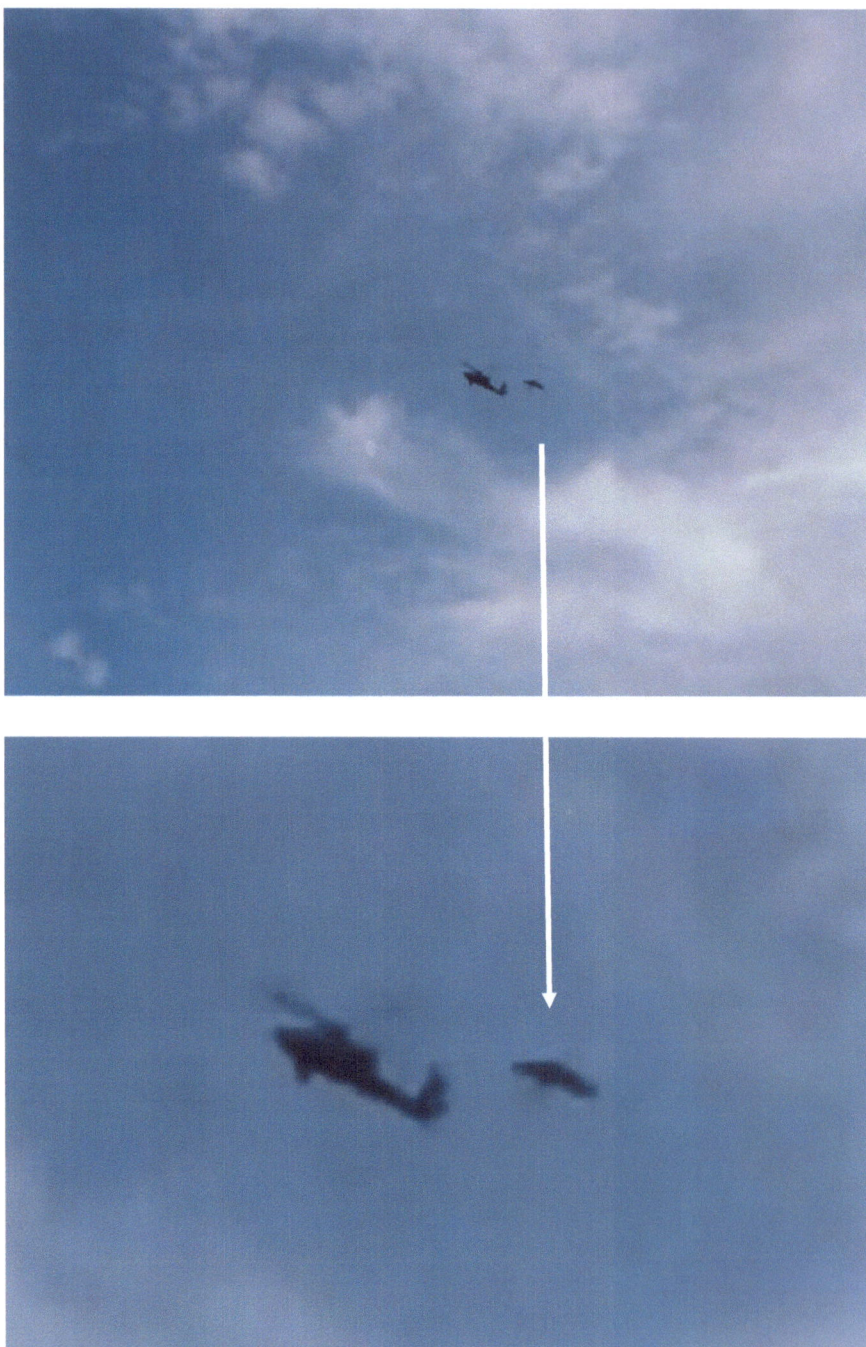

Photo and enlargement by Tony Degn

Photo #2 and enlargement of humanoid creature with wings by Tony Degn:

Thousands of Fast-Movers Captured on Video

Gary Stansbury used the Quantum UFO Observation Technique to capture a video of literally thousands of fast-movers in June of 2016. The objects move so fast that it is difficult to see them. Even at 1/10th speed they move very quickly. The full video is here: https://www.youtube.com/watch?v=3tILhsFKC9c&t=1s Following are a few stills from the video taken at 1 second intervals. Note the coordinated movement across the sky.

10:22

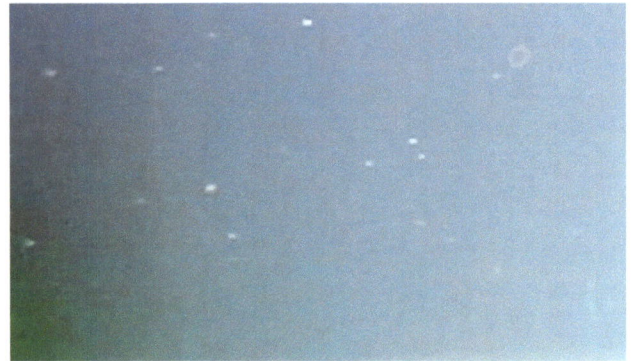

10:25

10:23

10:26

10:24

10:27

More still shots from Gary's video: A large craft appears in the lower center of frame 15:04, and by frame 15:05 it is almost out of view in the upper left corner.

Frame 15:04

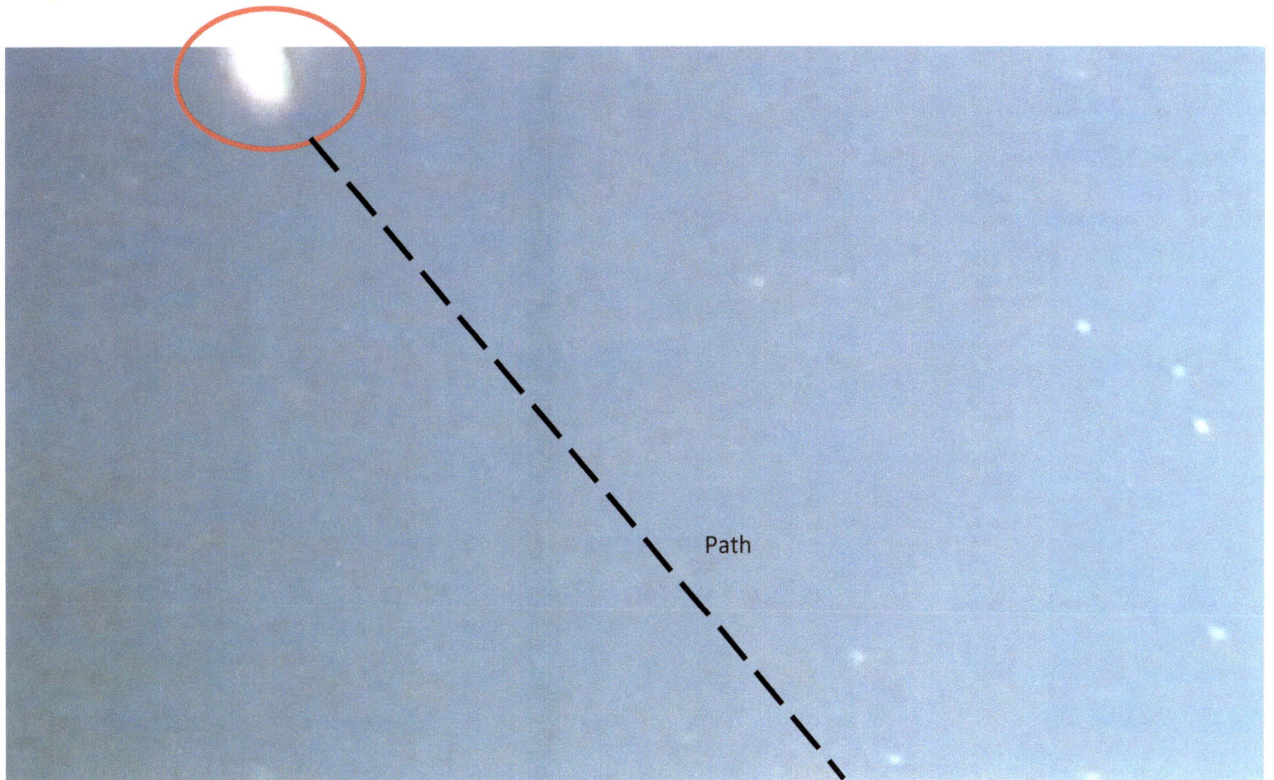

Path

Frame 15:05

Photos by NASA

The following stills were captured from a NASA video taken on the International Space Station on Feb. 20, 2020. The video lasts 22 minutes in which an object appears for the entire time. Wayne Lawrence discovered that there were some Fast Movers toward the end of the video as the station was further away. It makes one wonder how many images of Fast Movers NASA has on file. Scott C. Waring, the founder of UFO Sightings Daily, was the one who first spotted the unidentified flying object, which was later identified, and he posted a video showing exactly where it was and what it looks like. The large object is a Japanese Module, however, there are Fast Movers around it!

As Scott points out, whoever at NASA was controlling the camera noticed the objects as well because they zoomed in on it. No one from the space agency has commented yet on what it could be, but it isn't the first time something like this has happened. In 2016, a strange light was spotted near Earth on the NASA live feed. Soon after it appeared, the feed was cut. NASA later stated the object was either space junk, a reflection, or light from Earth.

Frame :56 seconds

Clip taken from NASA video. This video has been taken down from YouTube by NASA.

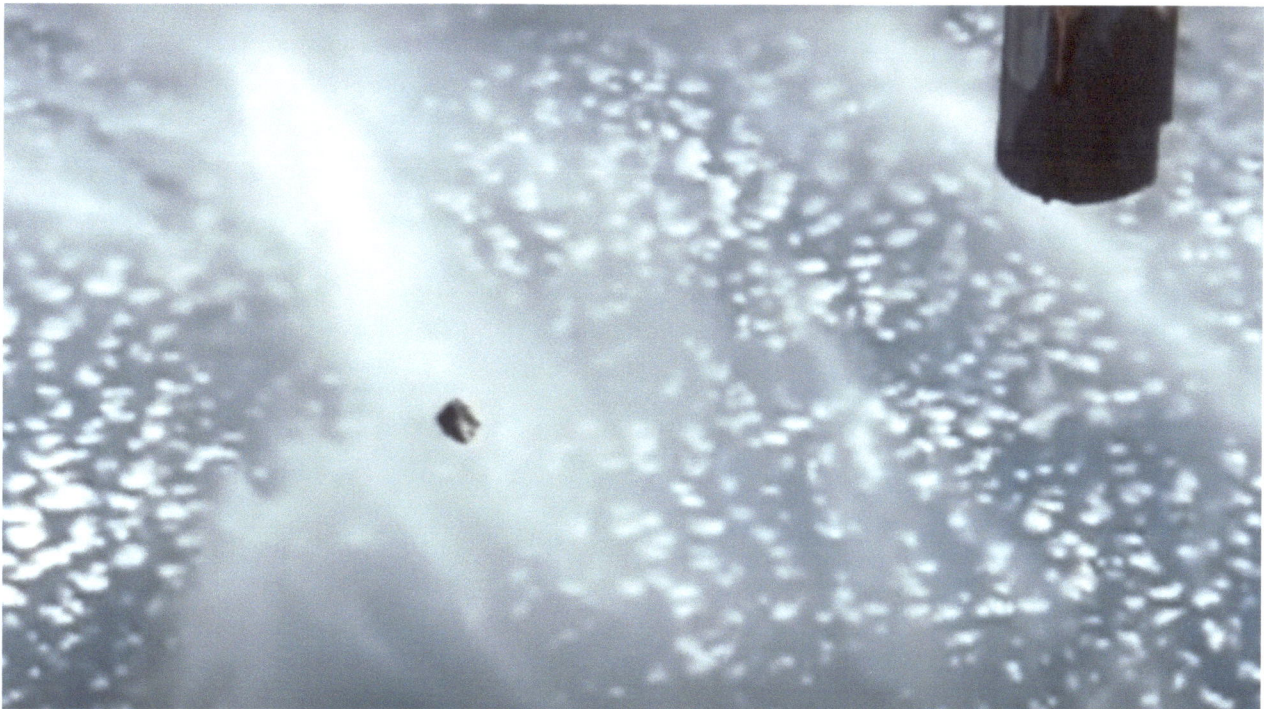

Frame 3:41

NASA has stated online that the object is actually part of a Japanese portion of the ISS that was jettisoned. It supposedly was the ICS-EF, or Inter-Orbit Communication System - Exposed Facility.

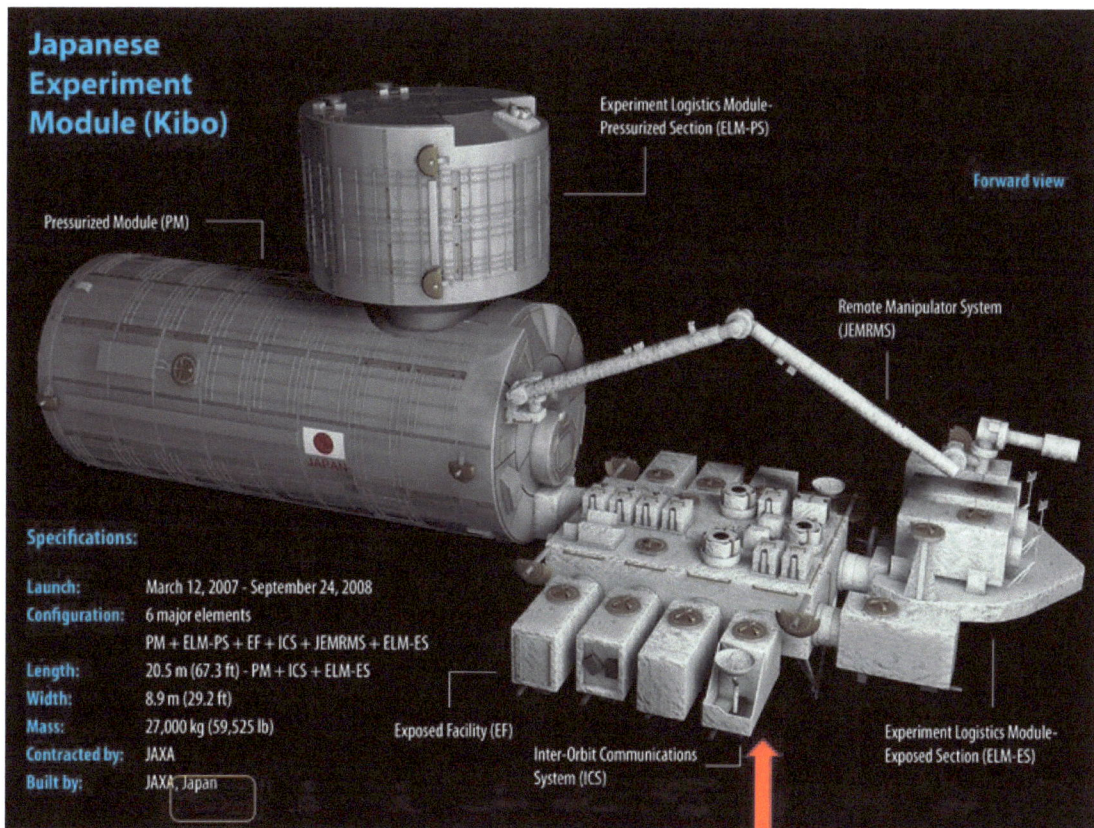

Another image of an object taken from the International Space Station. Right: close up photo.
Photo: NASA

Enhanced by Wayne Lawrence

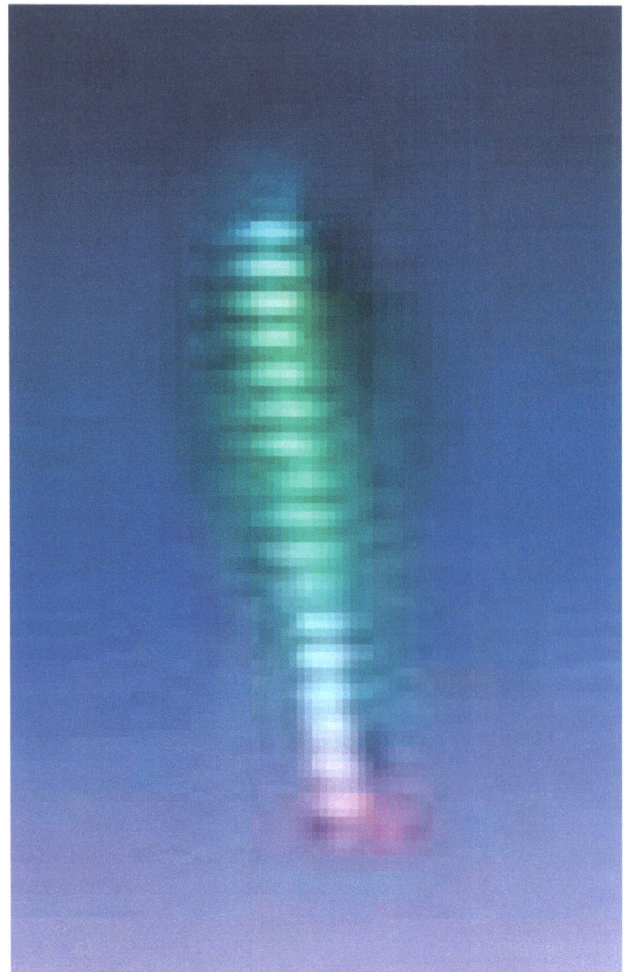

Frame 16:24

Frame 18:00 White light with Fast-Mover at the left

White light enhanced by Wayne Lawrence and UFO cropped below:

More NASA Images—The 12-mile Tether Incident

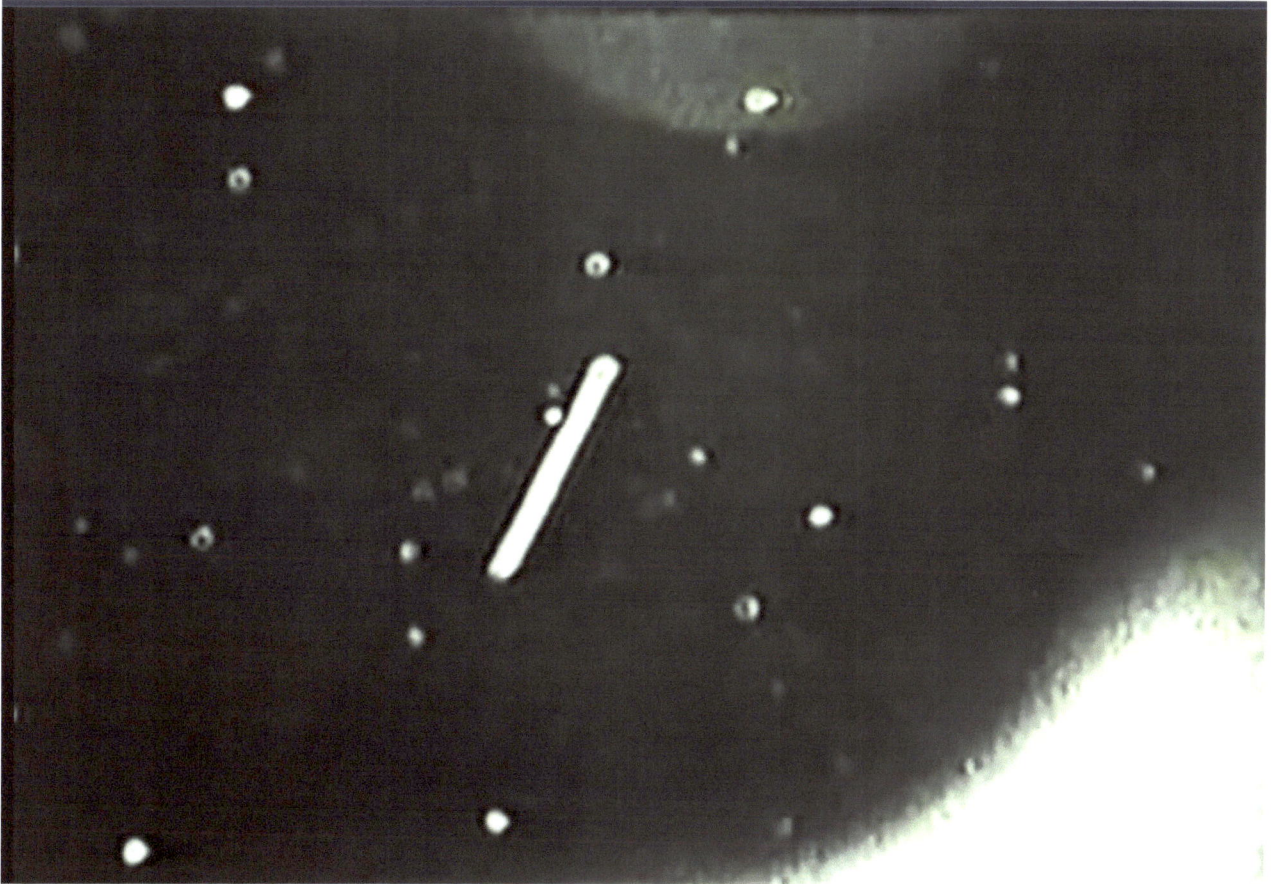

NASA image of the tether and unidentified flying objects around it

On February 25, 1996 NASA space shuttle mission STS-75 carried a large spherical satellite with a tether attached to the shuttle. The 12.5-mile tether was to be used for experimentation with electrical power generation from the ionosphere. The experiment was called the Tethered Satellite System, and it was to attempt to generate electricity by utilizing Earth's magnetic field. However, the tether accidentally broke away and what followed is intriguing to say the least.

The tether was swarmed with moving objects, many of which can be seen moving behind the tether indicating enormous size (1-3 miles in diameter). If you watch carefully you'll notice that some of the objects actually change direction, showing that they are under some sort of intelligent control. The fact that some of the objects move behind the tether, which is quite distant from the space shuttle, makes it apparent that the objects are not dust or debris on the camera lens or window.

Perhaps the UFOs are only visible due to the type of ultraviolet camera used on this mission. If you watch the video carefully, Fast-Movers can be seen as well.

Check YouTube for many videos of this incident. Following are stills taken from the video.

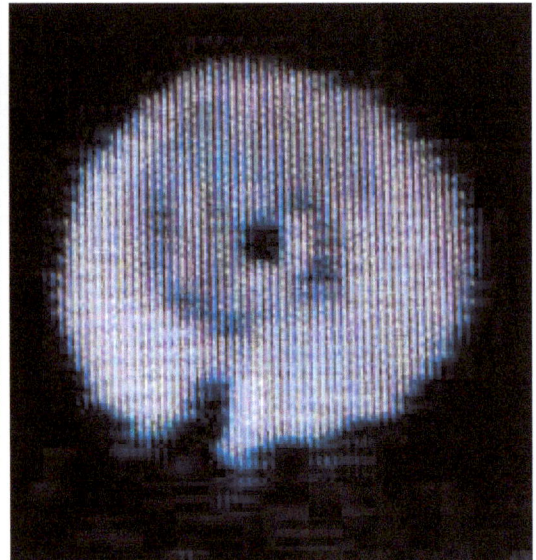

Close-up of one of the objects

Close-up of one of the larger craft shown behind the top of the tether.

Conversation between Houston and the Astronauts:

Astronaut Franklin: "You guys getting the image?"

Houston Control: "Franklin we see a long line, a couple of star-like things and a lot of things swimming in the foreground. Can you describe what you're seeing?"

Astronaut Franklin: "Well the long line is the tether and, uh, there's a little bit of debris that kind of flies with us and, uh, it's illuminated by the Sun at such a low angle so there's a lot of stray light and it's getting washed out quickly..."

The satellite (From NASA)

It is obvious from the conversation that something unusual was going on, but certainly we'll never hear from NASA that these were UFOs. Perhaps we are getting more images of UFOs or UAPs because of the improvements in our photographic equipment and the ability of these cameras to capture more objects that are just outside the visible light spectrum.

The Pentagon Officially Released the Navy UFO/UAP Videos on April 27, 2020

The general public was never supposed to see them, but we did in our skies for over a decade. In an unprecedented move, the Pentagon released three classified videos taken by US Navy pilots that appear to be unidentified flying objects. The Navy just put the three videos—titled "FLIR.mp4," "GOFAST.wmv," and "GIMBAL.wmv"—on its Freedom of Information Act (FOIA) page, a repository for documents released under the federal law that allows for the full or partial disclosure of U.S. government information to the public. The clips were first released in 2017 and 2018 by *The New York Times* and To The Stars Academy of Arts & Science, a UFO research group from former blink-182 member Tom DeLonge.

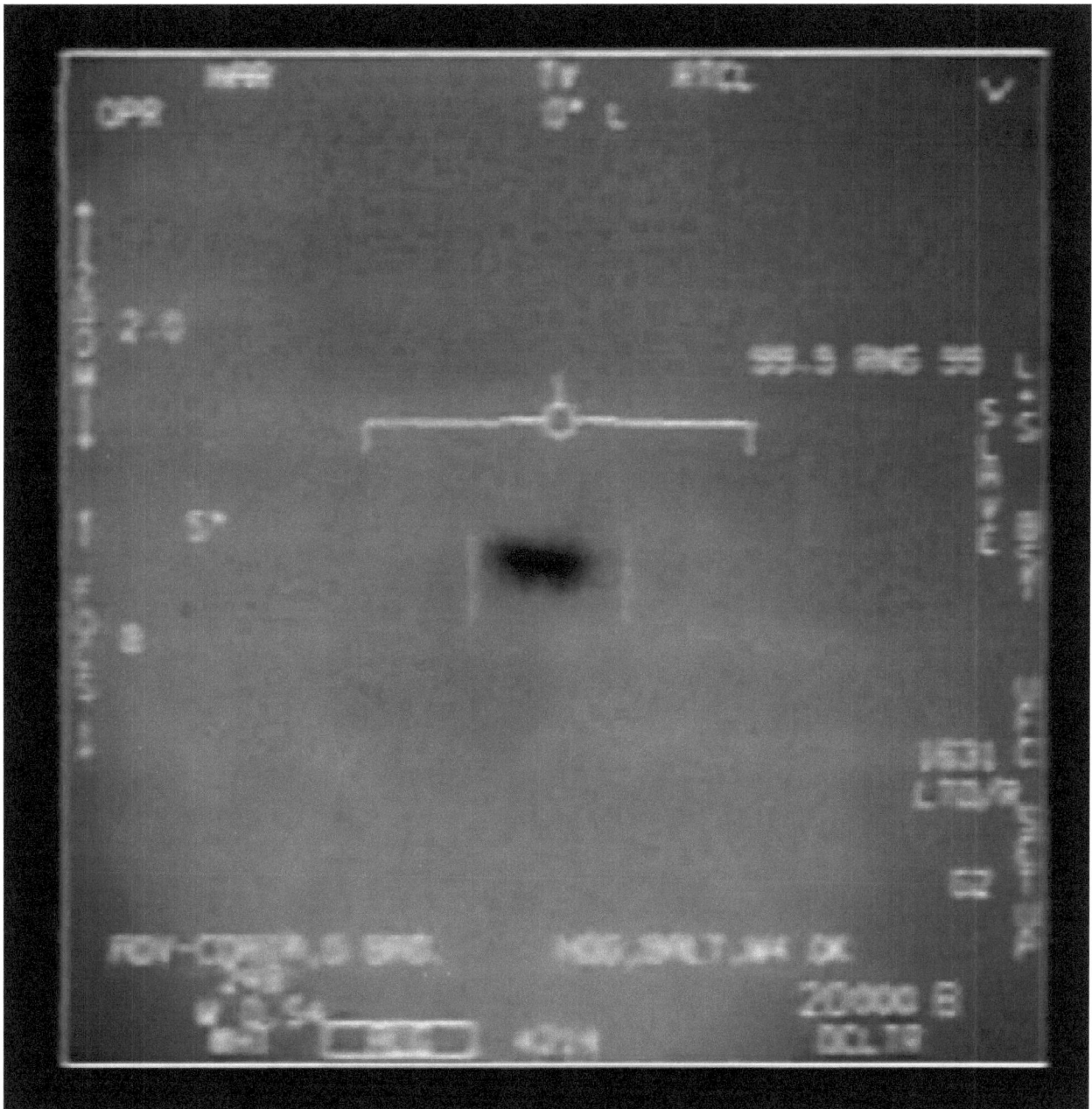

The infrared video was filmed in 2004 by two Navy fighter pilots with the Nimitz Carrier Strike Group 100 miles into the Pacific ocean. They captured a Tic-Tac shaped UFO "around 40 feet long and oval in shape.

The 2015 videos were filmed off of the east coast of the U.S. by an F/A-18F fighter jet using the aircraft's onboard Raytheon AN-ASQ-228 Advanced Targeting Forward-Looking Infrared pod. The video's were leaked in 2007 and 2017 without authorization. One of the pilots said of the 2004 video:

"As I got close to it ... it rapidly accelerated to the south, and disappeared in less than two seconds," retired US Navy pilot David Fravor told *CNN* in 2017. "This was extremely abrupt, like a ping pong ball, bouncing off a wall. It would hit and go the other way."

In one of the 2015 videos, a voice can be heard saying: "There's a whole fleet of them."

Another voice adds: "They're all going against the wind. The wind's 120 knots to the west. Look at that thing, dude!"

In an effort to "clear up any misconceptions by the public on whether or not the footage that has been circulating was real." the DoD, who had previous confirmed the videos were filmed by the US Navy, had made all three clips public.

The incredible speed of the UFOs indicates

Photos: US Navy

that they are Fast Movers. The fact that the videos were captured using FLIR infrared technology is an indication that the objects exist out of our range of sight using the naked eye, out side the visible light spectrum. Nevertheless, they do exist.

Watch the videos here: https://www.navair.navy.mil/foia/documents

The following is the most recent press release by defence.gov:

Immediate Release

Establishment of Unidentified Aerial Phenomena Task Force

Aug. 14, 2020

On Aug. 4, 2020, Deputy Secretary of Defense David L. Norquist approved the establishment of an Unidentified Aerial Phenomena (UAP) Task Force (UAPTF). The Department of the Navy, under the cognizance of the Office of the Under Secretary of Defense for Intelligence and Security, will lead the UAPTF.

The Department of Defense established the UAPTF to improve its understanding of, and gain insight into, the nature and origins of UAPs. The mission of the task force is to detect, analyze and catalog UAPs that could potentially pose a threat to U.S. national security.

As DOD has stated previously, the safety of our personnel and the security of our operations are of paramount concern. The Department of Defense and the military departments take any incursions by unauthorized aircraft into our training ranges or designated airspace very seriously and examine each report. This includes examinations of incursions that are initially reported as UAP when the observer cannot immediately identify what he or she is observing.

https://www.defense.gov/Newsroom/Releases/Release/Article/2314065/establishment-of-unidentified-aerial-phenomena-task-force/

We have never been alone and hope you will discover for yourself what you've been missing.

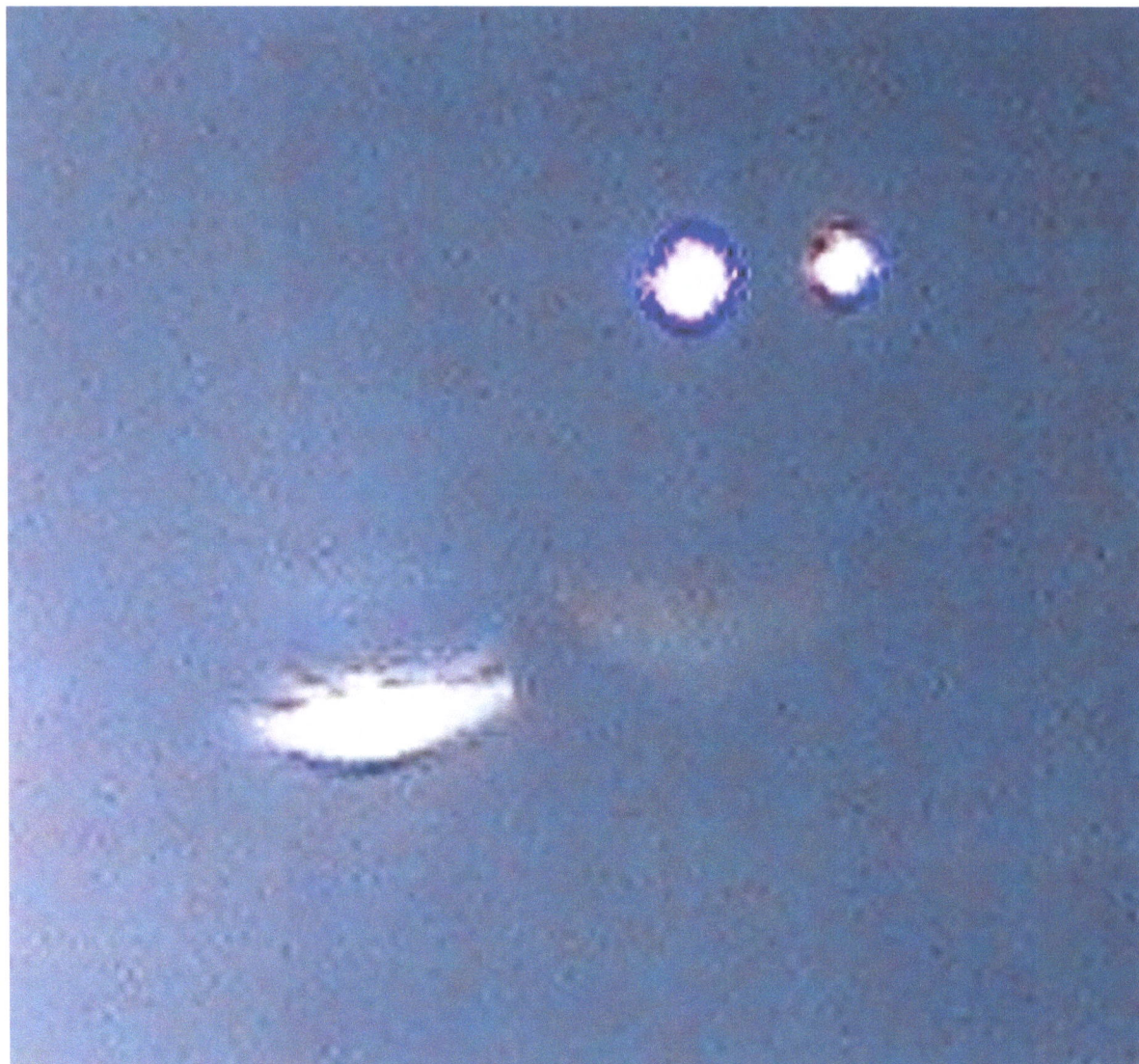

Photo of craft by Bill Spicer

THEY ARE HERE!

Now that you've seen our evidence for the existence of Fast-Movers, are you ready to try some of our techniques for viewing or capturing these objects on film? We'd love to hear from you. Write to Margie Kay at margiekay06@yahoo.com and send us your photos and video.

REFERENCES

Bill Spicer's YouTube Channel:
https://m.youtube.com/playlist?list=UUXyBTI78FOGhIhnQJ9XTfKA

Stranger at the Pentagon by Frank Stranges

Beyond Watching-A Journal of Experiences-Understanding UFO's by Bill Spicer

The Remote Viewing Workbook 2019 by Margie Kay www.margiekay.com

Confirmation by Whitley Strieber 2014

Kerry Walker with "Alice Eats the Apple" podcast interviews Bill Spicer:
https://www.youtube.com/channel/UCcb1O_LX4tYJ1PnF1AXwkww/videos

www.AliceEatsTheApple.com

Short video of UFOs captured Nov.3, 2019 by Bill Spicer using the Quantum UFO Observation Technique: https://www.youtube.com/watch?v=Ru1XbytayFk

Article by Margie Kay about the *Mass UFO Sighting in Kansas City June 20, 2019*:
www.missourimufon.com

Inflationary Vacuum State Propulsion Patent: https://patents.google.com/patent/US6960975B1/en

Interview with Boyd Bushman about Area 51: https://m.youtube.com/watch?v=VA3HV_gfq80

Immrama Insight CD's for meditation: www.inrama.org

Janalea Hoffman's CD's for meditation: www.rhythmicmedicine.com

The Space Tether Experiment: https://www-istp.gsfc.nasa.gov/Education/wtether.html

Gary Stansbury's You Tube Video: https://www.youtube.com/watch?v=3tILhsFKC9c&t=1s

Navy Videos: https://www.defense.gov/Newsroom/Releases/Release/Article/2165713/statement-by-the-department-of-defense-on-the-release-of-historical-navy-videos/

Double-Split Experiment: https://en.wikipedia.org/wiki/Double-slit_experiment
DOD Press release: https://www.defense.gov/Newsroom/Releases/Release/Article/2314065/establishment-of-unidentified-aerial-phenomena-task-force/

Adding a polarized filter to your iphone camera with filter adapter and camera lenses: https://youtu.be/8o7QNhdv-6E

ABOUT THE AUTHORS

Bill Spicer

Specializing as an aircraft engineering consultant, Bill works as a design and technical lead for a small aviation design company. He has an aviation career that spans 45 years. He graduated from Kansas State University with a degree in design and has worked on numerous aircraft, before specializing in the interior arrangements and certification.

He has authored and self published an academic textbook on aircraft interior design titled: "Introduction to Aircraft Interiors". In addition, he also holds a private pilot certificate, remote UAS (Drone-Part 107) pilot certificate and is an active aviation enthusiast. He is also an amateur astronomer and has made commercial videos using his UAS (Drone) for small businesses.

In 2010 he authored a book of his experiences about being able to see UFOs/UAPs during the daytime using the Quantum UFO Observaton Technique to record the UFOs/UAPs using a digital camera with proper UV filtering. He self published the book titled "Beyond Watching-A Journal of Experiences-Understanding UFOs", outlining his experience of how he came about this discovery.

During this process he communicated extensively with known astrophysicist Dr. Rudy Schild (Harvard) and others in the study of UFO's, such as author retired Rev. Barry Downing. His photos used in the book were examined by Dr. Rudy Schild and he stands by that they are not faked and the quantum physics knowledge observed and shared, may have proven a theory of how the UFO's operate. Through astral visions/dreams, he was shown how the UFO's use mental imagery, electrodynamics to propel the orbs/ships and can go from matter to energy and cross time-space using quantum physics.

His findings and observations were later published in the July/August, 2011 issue of UNX News Magazine, titled "New Technique Captures Daytime UFOs ".

Since 2016, he has made numerous presentations and demonstrations to small groups and guided private individuals over the world on how to see and record the UFOs/UAPs during the daytime. He also posts regularly on multi-media platforms his observations and encourages others to try the Solar Obliteration Technique.

Bill lives in Wichita, KS with his wife. Contact Bill at beyondwatching@gmail.com

Wayne Lawrence, MA

Wayne Lawrence was the Chief Investigator with the Missouri chapter of the Mutual UFO Network, and is now a field investigator in Illinois. He enjoys investigating the various sighting cases that he is assigned, as well as his own sightings. He also serves on the MUFON video analysis team and assists other investigators by examining video evidence and providing an analysis of the video.

Wayne has made numerous video lecture presentations that exhibit images and data from cases that he has investigated.

Lawrence noticed objects moving in and out of Blue Springs Lake several years ago, and has paid particular attention to the activity in this area.

Wayne has been a frequent presenter at local and state MUFON chapter meetings and paranormal conferences. His presentations include topics such as discoveries on the Moon and Mars, Solar visitors, "Fast Movers" and highlights of his own sightings.

Wayne has had a number of his investigatory cases highlighted in the national publication *"The MUFON Journal"* over the last several years. One recent case was included in *"UFO Cases of Interest, 2019 Edition"*.

Wayne grew up in French Polynesia where he learned three languages. He is currently fluent in five languages.

Lawrence attended seminary school, graduating with a Master's Degree and worked for 18 years as Director of International Translations for a global denomination.

He lives in Chicago, Illinois with his wife. Wayne has two married children and three grandsons.

Margie Kay

Margie Kay is a nationally acclaimed remote viewer and psychic medium who has helped solve over 60 missing person, theft, and homicide cases for law enforcement, private investigators, and individuals around the world. Her accuracy amazes audiences.

Kay owns and operates a fire investigation company and chimney contracting business in Kansas City, Missouri. She is a Missouri Licensed Private Investigator. Margie also owns a real estate investment company and vacation rental homes. She is the owner of UnX Media Publishing, and serves as chief editor.

Kay serves as Director of Quest Paranormal Investigation Group, and as Assistant State Director for Missouri MUFON. She has served on several national boards for various organizations.

Margie is the author of *Haunted Independence, Missouri; Gateway to the Dead: A Ghost Hunter's Field Guide; The Remote Viewing Workbook; The Kansas City UFO Flaps; A Sonoma County Phenomenon: Evidence for an Interdimensional Gateway;* and more. She has written over 250 articles for magazines, newsletters, and blogs of various genres.

Kay appeared on the Hangar 1 episode *"UFO Hot Spots"* and was featured on CNN and other news stations during the Kansas City Lights sightings, and was in the pilot TV episode for *Strange.* She has appeared on multiple radio programs including Jeff Rense, Jimmy Church, Joe Montaldo, Tracie Austin, Race Hobbs, and others.

Kay was the host of UNX News Radio on KGRA DB radio network for two years and hosted Quest Radio Show on KCXL 1140 AM in Kansas City for five years. She has spoken and presented at numerous meetings and conventions. She teaches Remote Viewing Workshops online and at conferences.

Margie studied acting, music and broadcasting at CMSU, UMC, and UMKC and worked as a professional bass player in several bands and orchestras. She played in the Northland Symphony, and many jazz and fusion groups.

Margie lives in Independence, Missouri with her husband, Geno and their cat, Patches.

Contact Margie at margiekay06@yahoo.com
Website: www.margiekay.com

UNXMEDIA

PUBLISHING

Publications by Un-X Media

Family Secrets 2017 by Jean Walker

Haunted Independence, Missouri 2017 by Margie Kay

Gateway to the Dead: A Ghost Hunter's Field Guide 2013 by Margie Kay

The Kansas City UFO Flaps 2017 by Margie Kay

The Remote Viewing Workbook 2019 by Margie Kay

A Sonoma County Phenomenon 2020 by Margie Kay

Un-X News Magazine 2001-2015 and 2001-2022

The Color Therapy Wall Chart 1999

Rules for Goddesses by Margie Kay 2003

The Fast Movers by Margie Kay, Bill Spicer, and Wayne Lawrence 2020

Doorway to Spirit by Devin Listrom 2020

The Alien Colonization of Earth's Waterways 2021

And coming soon:

The Master Dowser's Chart Book by Margie Kay 2022

THOR by Margie Kay 2022

UFO Hot Spot: Piedmont, Missouri 2022

www.unxmedia.com

Email: editor@unxmedia.com

All books available at Amazon.com and BarnesandNoble.com

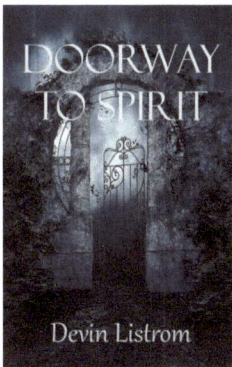

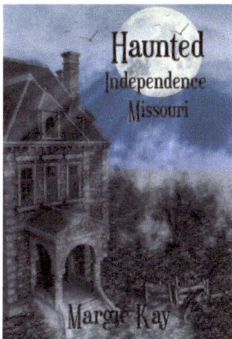

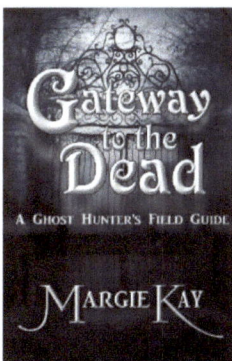

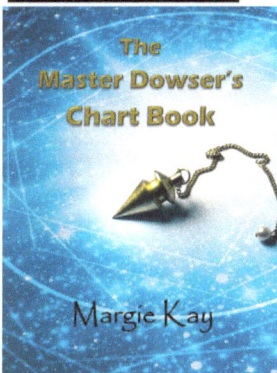

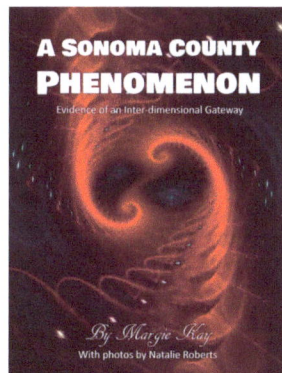

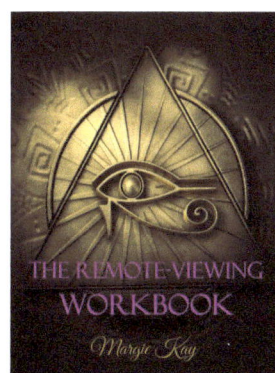

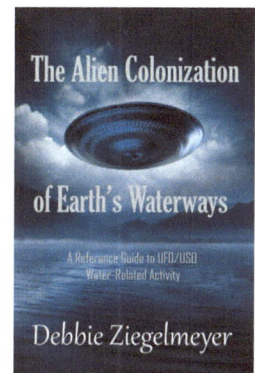

www.ingramcontent.com/pod-product-compliance
Lightning Source LLC
Chambersburg PA
CBHW060801270326
41926CB00002B/49